PROGRESS IN

Molecular Biology and Translational Science

Volume 89

PROGRESS IN Molecular Biology and Translational Science

G Protein-Coupled Receptors in Health and Disease, Part B

edited by

Ya-Xiong Tao

Department of Anatomy, Physiology and Pharmacology
College of Veterinary Medicine
Auburn University, Auburn
AL, USA

Volume 89

AMSTERDAM • BOSTON • HEIDELBERG • LONDON
NEW YORK • OXFORD • PARIS • SAN DIEGO
SAN FRANCISCO • SINGAPORE • SYDNEY • TOKYO
Academic Press is an imprint of Elsevier

Academic Press is an imprint of Elsevier
32 Jamestown Road, London, NW1 7BY, UK
Radarweg 29, PO Box 211, 1000 AE Amsterdam, The Netherlands
30 Corporate Drive, Suite 400, Burlington, MA 01803, USA
525 B Street, Suite 1900, San Diego, CA 92101-4495, USA

This book is printed on acid-free paper. ∞

Copyright © 2009, Elsevier Inc. All Rights Reserved

No part of this publication may be reproduced, stored in a retrieval system or transmitted in any form or by any means electronic, mechanical, photocopying, recording or otherwise without the prior written permission of the Publisher

Permissions may be sought directly from Elsevier's Science & Technology Rights Department in Oxford, UK: phone (+44) (0) 1865 843830; fax (+44) (0) 1865 853333; email: permissions@elsevier.com. Alternatively you can submit your request online by visiting the Elsevier web site at http://elsevier.com/locate/permissions, and selecting *Obtaining permission to use Elsevier material*

Notice
No responsibility is assumed by the publisher for any injury and/or damage to persons or property as a matter of products liability, negligence or otherwise, or from any use or operation of any methods, products, instructions or ideas contained in the material herein. Because of rapid advances in the medical sciences, in particular, independent verification of diagnoses and drug dosages should be made

Library of Congress Cataloging-in-Publication Data
A catalog record for this book is available from the Library of Congress

British Library Cataloguing in Publication Data
A catalogue record for this book is available from the British Library

ISBN: 978-0-12-374756-3
ISSN: 1877-1173

For information on all Academic Press publications
visit our website at elsevierdirect.com

Printed and Bound in the USA
09 10 11 12   10 9 8 7 6 5 4 3 2 1

Working together to grow
libraries in developing countries

www.elsevier.com | www.bookaid.org | www.sabre.org

ELSEVIER   BOOK AID International   Sabre Foundation

Contents

Contributors ix
Preface xi

GPR56 and Its Related Diseases 1
Zhaohui Jin, Rong Luo, and Xianhua Piao

I. GPR56 2
II. GPR56 and Brain Malformation 4
III. The Role of GPR56 in Brain Development 7
IV. GPR56 and Cancer 8
V. GPR56 Signaling 9
VI. Concluding Remarks 9
References 10

V2R Mutations and Nephrogenic Diabetes Insipidus ... 15
Daniel G. Bichet

I. Cellular Actions of Vasopressin 16
II. Rareness and Diversity of *AVPR2* Mutations 20
III. Most Mutant V2 Receptors Are Not Transported to the Cell Membrane and Are Retained in the Intracellular Compartments 22
IV. Nonpeptide Vasopressin Receptor Antagonists Act as Pharmacological Chaperones to Functionally Rescue Misfolded Mutant V2 Receptors Responsible for X-Linked NDI 24
V. Gain of Function of the Vasopressin V2 Receptor: Nephrogenic Syndrome of Inappropriate Antidiuresis 24
References 26

Calcium-Sensing Receptor and Associated Diseases ... 31
Geoffrey N. Hendy, Vito Guarnieri, and Lucie Canaff

I. Calcium Homeostasis 32
II. CASR and Diseases 33
III. CASR is a Family C GPCR 34
IV. Human CASR 34
V. Orthosteric Agonists 35

VI.	Allosteric Modifiers	36
VII.	Structure and Function	37
VIII.	Receptor Downregulation and Protein Kinase C	45
IX.	Receptor-Activity-Modifying Proteins and CASR Trafficking	45
X.	Ubiquitination and Conformational Checkpoint in CASR Processing	46
XI.	CASR and Overview of Signaling Pathways	46
XII.	CASR and the Parathyroid	48
XIII.	CASR and the Renal Tubule	50
XIV.	Disorders Associated with CASR (Table I)	50
XV.	CASR Mutation Repertoire	54
XVI.	Autoantibodies and the CASR	66
XVII.	CASR Polymorphisms	66
XVIII.	Altered Expression of CASR and Disease	71
XIX.	CASR Allosteric Modifiers in the Clinic	73
XX.	Summary	73
	References	74

Diseases Associated with Mutations of the Human Lutropin Receptor 97

Deborah L. Segaloff

I.	Introduction	97
II.	The LHCGR and Human Physiology	98
III.	The LHCGR Protein and the *LHCGR* Gene	99
IV.	Activating Mutations of the LHCGR	101
V.	Inactivating Mutations of the *LHCGR*	103
	References	108

Follicle Stimulating Hormone Receptor Mutations and Reproductive Disorders 115

Ya-Xiong Tao and Deborah L. Segaloff

I.	Introduction	116
II.	Follicle Stimulating Hormone Receptor	116
III.	Inactivating *FSHR* Mutations and Hypergonadotropic Hypogonadism	118
IV.	Gain-of-Function *FSHR* Mutations and Spontaneous Ovarian Hyperstimulation Syndrome	121
V.	Structure–Function Insights from Studies of Constitutively Active FSHR Mutants	123
VI.	Conclusions	125
	References	125

The Human Prostacyclin Receptor: From Structure Function to Disease............ 133

Kathleen A. Martin, Scott Gleim, Larkin Elderon, Kristina Fetalvero, and John Hwa

I. History............	134
II. Molecular and Structural Biology............	138
III. Pathophysiology............	145
IV. Therapeutics............	152
V. Genetic Variants............	154
References............	157
Index............	167

Contributors

Numbers in parentheses indicate the pages on which the authors' contributions begin.

Daniel G. Bichet, Canada Research Chair in Genetics of Renal Diseases, Groupe d'Étude des Protéines Membranaires; and Department of Medicine and Physiology, Université de Montréal, Research Center; and Nephrology Service, Hôpital du Sacré-Coeur de Montréal, Montréal, Québec, Canada H4J 1C5 (15)

Lucie Canaff, Departments of Medicine, Physiology, and Human Genetics, McGill University and Calcium Research Laboratory and Hormones and Cancer Research Unit, Royal Victoria Hospital, Montreal, Quebec, Canada H3A 1A1 (31)

Larkin Elderon, Department of Pharmacology and Toxicology, Dartmouth Medical School, Hanover, New Hampshire 03755 (133)

Kristina Fetalvero, Department of Pharmacology and Toxicology, Dartmouth Medical School, Hanover, New Hampshire 03755; and Department of Surgery, Section of Vascular Surgery, Dartmouth Hitchcock Medical Center, Lebanon, New Hampshire 03756 (133)

Scott Gleim, Department of Pharmacology and Toxicology, Dartmouth Medical School, Hanover, New Hampshire 03755 (133)

Vito Guarnieri, Medical Genetics Service, Hospital "Casa Sollievo della Sofferenza," Instituto di Recovero e Cura a Carattere Scientifico, San Giovanni Rotondo (Foggia), Italy (31)

Geoffrey N. Hendy, Departments of Medicine, Physiology, and Human Genetics, McGill University and Calcium Research Laboratory and Hormones and Cancer Research Unit, Royal Victoria Hospital, Montreal, Quebec, Canada H3A 1A1 (31)

John Hwa, Department of Pharmacology and Toxicology, Dartmouth Medical School, Hanover, New Hampshire 03755; and Department of Medicine, Section of Cardiology, Dartmouth Hitchcock Medical Center, Lebanon, New Hampshire 03756 (133)

Zhaohui Jin, Division of Newborn Medicine, Children's Hospital Boston, Harvard Medical School, Boston, Massachusetts 02115 (1)

Rong Luo, Division of Newborn Medicine, Children's Hospital Boston, Harvard Medical School, Boston, Massachusetts 02115 (1)

Kathleen A. Martin, Department of Pharmacology and Toxicology, Dartmouth Medical School, Hanover, New Hampshire 03755; and Department of Surgery, Section of Vascular Surgery, Dartmouth Hitchcock Medical Center, Lebanon, New Hampshire 03756 (133)

Xianhua Piao, Division of Newborn Medicine, Children's Hospital Boston, Harvard Medical School, Boston, Massachusetts 02115 (1)

Deborah L. Segaloff, Department of Molecular Physiology and Biophysics, The Roy J. and Lucille A. Carver College of Medicine, The University of Iowa, Iowa City, Iowa 52242 (97, 115)

Ya-Xiong Tao, Department of Anatomy, Physiology and Pharmacology, College of Veterinary Medicine, 212 Greene Hall, Auburn University, Auburn, Alabama 36849 (115)

Preface

G protein-coupled receptors (GPCRs) comprise the largest family of membrane proteins with 800 members. They transduce signals from a diverse array of endogenous ligands, including ions, amino acids, nucleotides, lipids, peptides, and large glycoprotein hormones. They are also responsible for our sensing of exogenous stimuli including photons and odorants. GPCRs regulate almost every aspect of our physiological functions. It is estimated that 40–50% of currently used therapeutic drugs target GPCRs directly or indirectly. Because the current drugs target only a small portion of the GPCRs, opportunities for targeting the remaining GPCRs are enormous.

As GPCRs are such versatile regulators of physiological processes, it is not difficult to imagine that mutations in these genes will result in dysregulation of these physiological processes, and therefore be of pathophysiological significance. The first naturally occurring mutation in a GPCR was identified in rhodopsin (causing retinitis pigmentosa) in 1990. Since then, many GPCRs were found to be mutated in a plethora of diseases. Several volumes as well as numerous review articles were published on this topic. The current volumes are updates on this rapidly advancing field. Leading experts present their accounts on a diverse array of these GPCRs in health and disease. These include some of the earlier classical examples of diseased GPCRs such as rhodopsin and V2 vasopressin receptor to more recent additions including the melanocortin-4 receptor, GPR54, and GPR56.

The studies of these naturally occurring mutations in GPCR genes have led to significant advances in our understanding of the physiology and pathophysiology of these GPCRs. Detailed understanding of the molecular defects are the basis of future personalized medicine. Some clinical trials have been done with impressive results. I hope the readers are excited to read the rapid progresses described in these two volumes. Indeed, these chapters are only selected examples of the field that combines thorough clinical observation, molecular genetics, pharmacology, biochemistry, cell biology, and epidemiology.

I thank all the authors for taking time out of their busy schedules to write their outstanding contributions. I apologize for the occasional nagging reminding the due dates of their chapters. I cherish the opportunity to get to know these leading scientists by emails. I look forward to meeting them in person at scientific meetings.

It has been a wonderful experience from the planning of the chapters and authors to seeing the chapters in print. For this, I am very grateful to Dr. P. Michael Conn, the Series Editor, for his trust, guidance, and friendship. I also thank the Editors at Elsevier, Ms. Lisa Tickner and Ms. Delsy Retchagar, for making these volumes a reality. It was a pleasure to work with them.

Finally, a special thanks to my wife, Zhen-Fang, and my daughters Nancy, Rachel, and Lily, for their understanding and love.

YA-XIONG TAO
Auburn, Alabama

GPR56 and Its Related Diseases

Zhaohui Jin, Rong Luo, and Xianhua Piao

Division of Newborn Medicine, Children's Hospital Boston, Harvard Medical School, Boston, Massachusetts 02115

I. GPR56 2
   A. A Member of Adhesion GPCRs 2
   B. GPS Motif 2
   C. The Biochemical Properties of GPR56 3
II. GPR56 and Brain Malformation 4
   A. Bilateral Frontoparietal Polymicrogyria 4
   B. Cobblestone-Like Cortical Malformation 5
   C. GPR56 Mutations Identified 5
III. The Role of GPR56 in Brain Development 7
   A. The Pial Basement Membrane 7
   B. GPR56 and the Pial BM 8
IV. GPR56 and Cancer 8
   A. A Suppressor of Tumor Progression 8
   B. The Oncogenic Property of GPR56 8
V. GPR56 Signaling 9
VI. Concluding Remarks 9
   References 10

GPR56, an orphan G protein-coupled receptor (GPCR), was identified one decade ago by two independent groups through a degenerate PCR-based approach for secretin-like GPCRs and by differential display of melanoma cell lines with different metastatic potentials. The finding that GPR56 was significantly downregulated in high metastatic melanoma cell lines suggests its possible role in cancer metastasis. However, the function of GPR56 remained unclear until 2004 when mutations in the human *GPR56* gene were found to cause a specific brain malformation called bilateral frontoparietal polymicrogyria. Although GPR56 is expressed in a wide range of tissues, the consequences of loss-of-function mutations in the *GPR56* gene have only been observed in the central nervous system. Studies from knockout mouse model indicate that GPR56 regulates brain development by affecting the integrity of the pial basement membrane in the developing brain.

I. GPR56

The human *GPR56* gene spans 45 kb at the genomic level and is composed of 14 exons with a coding region of 2061 bp (GenBank accession number AF106858) from exon 2 to 14. Northern blot analysis revealed that the *GPR56* gene was expressed in a wide range of tissues including brain, heart, thyroid, kidney, testis, pancreas, and skeletal muscle.[1] *In situ* hybridization on adult rat brain showed *GPR56* transcripts in the hippocampus, thalamus, hypothalamus, amygdala, and cortex.[1] In the developing mouse brain, *GPR56* mRNA is highly expressed in the embryonic ventricular zone, a region rich in neuronal progenitor cells.[2]

A. A Member of Adhesion GPCRs

Mammalian G protein-coupled receptors (GPCRs) are classified into three major groups based on their sequence similarity: rhodopsin (class A), secretin receptor (class B), and metabotropic receptor (class C).[3,4] Within the secretin-like class B subfamily, a subgroup of GPCRs forms its own phylogenetic cluster.[5] This subclass of receptors shares significant homology with other class B GPCRs across the 7-transmembrane region, but has an unusually long N-terminal extracellular domain. It is named as *l*ong *N*-terminal class B *7-t*rans*m*embrane proteins (LNB-7TM).[6] This subclass comprises 33 members in human genome. Their N-terminal segments often contain domains found in adhesion proteins, such as cadherin, lectin, immunoglobulin, and thrombospondin domains. Thus, they are also referred to as "adhesion GPCRs."

GPR56 is a member of the adhesion GPCR family. Of the 33 adhesion GPCRs identified, GPR56 protein is most closely related to six other adhesion GPCRs: GPR97, GPR112, GPR114, GPR126, GPR128, and HE6.[5] As seen in other members of adhesion GPCRs, the N-terminal segment of GPR56 contains a high percentage of serine and threonine residues and a GPCR *p*roteo*l*ytic *s*ite (GPS) motif just before the first transmembrane spanning domain.[5–10] There is no identifiable domain except for the GPS motif in GPR56.

B. GPS Motif

The GPS motif is a conserved cysteine-rich domain in the membrane-proximal region. The four cysteine and two tryptophan residues are arranged in a conserved sequence as $C\text{-}x_2\text{-}W\text{-}x_{6-16}\text{-}W\text{-}x_4\text{-}C\text{-}x_{11}\text{-}C\text{-}x\text{-}C$. A posttranslational proteolytic cleavage event within the GPS motif has been linked to the generation of heterodimeric receptors composed of an extracellular α-subunit and a 7TM β-subunit. The GPS domain was first demonstrated to be an internal cleavage site in latrophilin and was later found in many other adhesion GPCRs.[5–11] Several other proteins outside the adhesion GPCR family also contain the conserved GPS motif. These include human polycystic kidney disease protein 1 (PDK-1), a channel-like 11-span transmembrane protein suREJ3, and PKDREJ, the human homologue of suREJ3.[12–14] Although the functional significance of the GPS motif-associated proteolysis remains unknown, current literature supports the notion that it is important for the intracellular trafficking and cell surface expression of cell surface receptors.[10,15]

The GPS-mediated protein cleavage is thought to be an autocatalytic process.[16] The conserved four cysteine and two tryptophan residues are likely to be essential for the proper formation of the catalytic core. Two disease-associated mutations in the GPS domain of GPR56 substitute the conserved cysteine or tryptophan residues and thus disrupt this cleavage, resulting in uncleaved products and improper protein localization.[15] In addition to the cysteine and tryptophan, a highly conserved amino acid sequence in the GPS domain, FAVLM, is required for GPS-mediated proteolytic cleavage. The FAVLM sequence may also facilitate the association between N- and C-terminal subunits.[16,17]

C. The Biochemical Properties of GPR56

GPR56 is cleaved into an N- and a C-terminal fragment (GPR56^N and GPR56^C, respectively) in a manner consistent with GPS-mediated proteolysis.[15,18,19] The speculated cleavage site in GPR56 is between amino acids Histidine-381 and Leucine-382.[18,19] GPR56^N remains associated with GPR56^C noncovalently, forming heterodimers on the cell membrane that function as signaling complexes, as seen in several adhesion GPCRs.[15] Some GPR56^N are also secreted to the conditioned media.[15] It is possible that the secreted GPR56^N could diffuse locally to exert its biological function.

GPR56 is heavily glycosylated. There is a serine/threonine-rich region in the N-terminal extracellular portion of GPR56, suggesting the presence of possible O- and/or N-glycosylation sites. Indeed, GPR56^N was detected as a protein ladder ranging 60–80 kDa on Western blot analysis, which is much bigger than the predicted unmodified size of 40 kDa.[15] Proteins can be glycosylated at the amide nitrogen in asparagines (Asn) (N-linked) or at the oxygen in serine and threonine side chains (O-linked). Enzymatic analysis revealed that GPR56^N does not contain any O-linked sugars.[15] There are seven N-linked glycosylation consensus sites (Asn-Xaa-Ser/Thr) within GPR56^N at amino acid positions Asn-39, -148, -171, -234, -303, -324, and -341. Combination of site-direct mutagenesis and Western blot analysis confirmed that GPR56 is N-glycosylated at these seven sites. The biological significance of the N-glycosylations is not clear, although current literature suggests that it plays an important role for GPR56 protein intracellular trafficking and cell surface expression.[15]

II. GPR56 and Brain Malformation

A. Bilateral Frontoparietal Polymicrogyria

Polymicrogyria is malformation of the cerebral cortical development in which the brain surface is irregular and the normally convoluted gyri are replaced by numerous (poly) smaller (micro) gyri. Bilateral frontoparietal polymicrogyria (BFPP) is a recessively inherited genetic disorder of human brain development, caused by mutations in the *GPR56* gene.[2,20–22] Individuals with mutations in the *GPR56* gene share considerable clinical homogeneity with five common clinical features and three identified typical magnetic resonance imaging (MRI) findings.[2,20–22]

(1) *Mental retardation of a moderate to severe degree*—Mental retardation is evident in all BFPP patients. Most of them only have very limited verbal language and need help in simple daily activities like dressing. They usually are not toilet trained or are trained at a much later age.

(2) *Motor developmental delay*—Global motor developmental delay is usually evident during the first year of life. Developmental milestones are achieved at much later ages.

(3) *Seizures*—These are reported in 95% of BFPP patients. Most patients appeared to suffer from symptomatic generalized epilepsy. Seizures usually begin before 4 years of age. Seizure types, which vary among patients, include tonic, atonic, atypical absence, and myoclonic. Recently, Dr. Guerrini's group reported four BFPP cases associated with Lennox–Gastaut syndrome, characterized by a severe form of generalized seizure of more than one type.[23] A specific EEG abnormality called a slow spike-and-wave pattern is found when the child is awake, along with generalized fast rhythms while asleep.

(4) *Cerebellar signs*—Ninety-four percent of a cohort of 29 BFPP patients in a genotype–phenotype analysis reported to have cerebellar signs, consisting of truncal ataxia, finger dysmetria, and rest tremor.

(5) *Dysconjugate gaze*—This occurs in about 88% of reported cases in which the patient's gaze is described as esotropia, nystagmus, exotropia, strabismus, or a known history of a "squint."

(6) *Bilateral polymicrogyria*—Abnormally thickened cortex is evident in the entire cortex, with the frontal and parietal lobes most severely affected.

(7) *White matter defect*—Bilateral patchy white matter signal changes without specific pattern are evident in all BFPP patients.

(8) *Brainstem and cerebellar hypoplasia*—The brain stem and vermis are frequently small in BFPP patients.

The incidence of BFPP is unknown. This is largely due to the difficulties in making the diagnosis prior to the high-resolution MRI and molecular testing. Indeed, the confirmed 29 BFPP cases were originally reported under five different diagnoses: "autosomal recessive syndrome of pachygyria," "neuronal migration abnormality," "cobblestone lissencephaly," "lissencephaly with cerebellar hypoplasia," and "bilateral frontoparietal polymicrogyria." The life expectancy is expected to be shortened for BFPP patients when compared to the normal population due to chronic disabilities. The oldest reported BFPP patient was 29 years of age. One patient in the cohort of 29 reported BFPP cases died of infection at age 4.[22]

B. Cobblestone-Like Cortical Malformation

Genotype–phenotype analysis in patients with BFPP and other similar polymicrogyria syndromes have demonstrated that *GPR56* sequence alterations define a characteristic clinical syndrome similar to cobblestone-like cortical malformation.[22] Studies in *Gpr56* knockout mice revealed a classic cobblestone-like cortical phenotype, confirming the histopathology of BFPP is cobblestone cortical malformation.[24]

Cobblestone cortex results from aberrant neuronal migration through the breaches in the basal lamina.[25] Cobblestone cortex is typically seen in three distinct human disorders: muscle–eye–brain (MEB) disease, Fukuyama-type congenital muscular dystrophy (FCMD), and Walker–Warburg syndrome (WWS).[25] These three disorders are autosomal recessive diseases that encompass congenital muscular dystrophy, ocular malformations, and cobblestone lissencephaly. MEB, FCMD, and some WWS cases are caused by aberrant glycosylation of α-dystropglycan, a receptor for laminin.[26–28] Mutant mice with deletions in some members of integrin pathway molecules and the extracellular matrix (ECM) constituents also show cortical migration defects with deficiencies in basal lamina integrity and cortical ectopia, features that resemble human cobblestone malformation.[29–34] To date, all of the cobblestone cortex-causative genes identified encode proteins that are directly or indirectly involved in basal lamina assembly.[26–31]

C. GPR56 Mutations Identified

To date, a total of 13 distinct GPR56 mutations have been reported in BFPP patients, including 1 deletion, 2 splicing, and 10 missense mutations (Table I and Fig. 1).[22,23] The deletion mutation, splicing mutations, and all missense mutations except one give rise to an indistinguishable BFPP phenotype, suggesting that they are probably all null mutations. However, the C346S mutation also causes an excessively small brain called microcephaly in addition

TABLE I
Mutations in *GPR*56 Associated with BFPP

Nucleotide change[a]	Exon/intron	Predicted protein	Clinical phenotype	References
97C > G	2	R33P	BFPP + Lennox–Gastaut syndrome	23
112C > T	3	R38W	BFPP	22
113G > A	3	R38Q	BFPP	22
235C > T	3	R79X	BFPP + Lennox–Gastaut syndrome	23
263A > G	3	Y88C	BFPP	22
272G > C	3	C91S	BFPP	22
739–746 delCAGGACC	5	Frame shift[b]	BFPP	22
E5-1G > C	5	Splicing mutation	BFPP	22
IVS9+3G > C	Intron 9	Splicing mutation	BFPP	22
1036T > A	8	C346S	BFPP + microcephaly	22
1046G > C	8	W349S	BFPP	22
1693C > T	13	R565W	BFPP + Lennox–Gastaut syndrome	22, 23
1919T > G	13	L640R	BFPP	22

[a]Base pair counted from starting codon ATG. IVS, intron; E, exon; (+) denotes intronic position 3′ of splice junction in donor and (−) denotes exonic position 5′ of splice junction in donor. For example, IVS9 + 3 means 3 bases of 3′ of the splice donor junction of intron 9.

[b]Deletion of 7 bp that alters the translational reading frame, resulting in truncated protein with premature protein termination.

to BFPP, a more severe phenotype than all the other mutations identified.[2,22] Further investigation of the pathophysiology associated with C346S mutation will likely shed light on their molecular mechanism associated with different clinical manifestations.

The four missense mutations in the tip of GPR56^N (R38Q, R38W, Y88C, and C91S) produce proteins with reduced intracellular trafficking and poor cell surface expression. The two mutations in the GPS domain (C346S and W349S) produce proteins with dramatically impaired cleavage that fail to traffic beyond the endoplasmic reticulum. Chemical chaperones such as thapsigargin and 4-phenylbutyrate can partially rescue mutant GPR56 cell surface expression in cells expressing the mutant receptors (R38Q, R38W, Y88C, C91S, C346S, or W349S).[15]

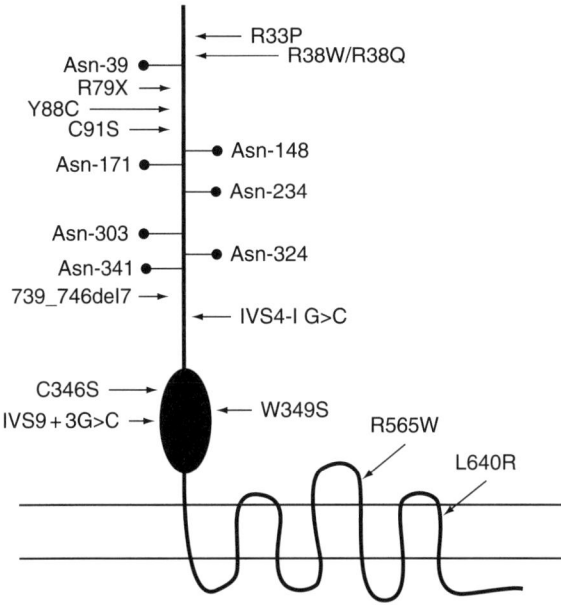

FIG. 1. Schematic representation of GPR56 with the known protein domain and mutations identified in BFPP patients. Arrows indicate positions for BFPP-associated mutations. There is a total of ten missense mutations, six at the tip of the N-terminus (R33P, R38Q, R38W, R79X, Y88C, and C91S), two in the GPS domain (C346S and W349S), one in the second extracellular loop (R565W), and one in the last transmembrane spanning domain (L640R). ●, GPS domain; ●——, site for N-glycosylation.

III. The Role of GPR56 in Brain Development

A. The Pial Basement Membrane

The pial basement membrane (BM) is a specialized ECM that overlies the surface of the developing brain. Microscopically, the pial BM is formed adjacent to the radial glial endfeet. Meningeal fibroblasts contribute to the pial BM by secreting and organizing the majority of basal lamina constituents, including laminin, collagen IV, nidogen, and the heparin sulfate proteoglycan.[35]

Radial glia cells are special progenitor cells in the developing cortex. They have their somata in the ventricular zone and extend their long radial processes through the entire cortex, attaching via their endfeet to the pial BM. The proper anchorage of the radial glial endfeet is highly relevant to the integrity of the pial BM.[29–33,36–38]

B. GPR56 and the Pial BM

Studies on human and mouse models of cobblestone-like malformations have identified laminin receptors, α-dystroglycan, and some integrin family members, as molecules serving critical roles in pial BM assembly/maintenance.[29–33,36–38] Deletion of α3β1 and α6β1 integrins results in severe disruption of pial BM and cortical lamination defects, indicating a role for laminin–integrin interactions in the maintenance of the pial BM and in cortical development.[30] We have demonstrated that GPR56 functions as a new ECM receptor, whose ligand in the pial BM is yet unknown. It remains elusive whether GPR56 interacts with integrins in regulating the pial BM integrity.

IV. GPR56 and Cancer

A. A Suppressor of Tumor Progression

The role of GPR56 in cancer metastasis is best established in melanoma. *GPR56* mRNA is found to be significantly downregulated in several highly metastatic melanoma cell lines compared with poorly metastatic cells, suggesting its role in cancer metastasis.[39] Subsequently, Dr. Hynes' group extended this observation by demonstrating the following findings:[18]

(1) GPR56 is downregulated in highly metastatic melanoma cells.
(2) Overexpression of GPR56 in high metastatic melanoma cells suppressed tumor growth and their metastatic ability, while downregulation of GPR56 by siRNA enhanced tumor growth and metastasis.
(3) The growth/metastasis inhibition by GPR56 only occurs *in vivo*, suggesting a need for a specific factor in the tumor microenvironment.
(4) Tissue transglutaminase (TG2), a ubiquitously expressed crosslinking enzyme, is the specific factor that binds to GPR56^N. TG2 itself has been shown to play suppressive roles in tumor progression.[40–43] Interactions between GPR56 and TG2 may facilitate the establishment of a hostile microenvironment to inhibit metastasis.

B. The Oncogenic Property of GPR56

Foehr's group first reported upregulation of GPR56 in aggressive form of brain tumors, glioblastoma and astrocytoma, at mRNA and protein levels.[19] Overexpression of GPR56 was later reported in many different cancer tissues, including ovarian cancer, pancreatic cancer, colon cancer, nonsmall cell lung cancer, and esophageal cancer compared with their normal counterparts.[44] Reduction of GPR56 by siRNA significantly induced apoptosis and reduced anchorage-independent growth of cancer cells *in vitro*, and reduced tumor size *in vivo*.[44] Further mechanistic studies suggest that GPR56 likely exerts its oncogenic property by affecting cell adhesion and/or migration.[19,44]

The above reports indicate the role of GPR56 in tumor cell biology. Further study is needed to elucidate the role of GPR56 in cancer development and progression. It would also be interesting to investigate the cancer incidence in individuals harboring mutations in the *GPR56* gene.

V. GPR56 Signaling

The signaling properties of GPR56 are poorly understood. GPR56 was shown to be associated with the membrane-bound tetraspanin molecules CD9 and CD81 *in vitro*, and CD9/CD81/GPR56 complexes facilitate association with $G\alpha_{q/11}$ and $G\beta$ subunits.[45] The function of this tetraspanin–GPR56 complex remains unclear. Members of the tetraspanin family of cell surface proteins act as molecular scaffolds with the adhesion proteins such as integrins to facilitate their functions.[46]

It has been suggested that GPR56 regulates cell migration by inducing transcription activation and actin fiber reorganization.[19,47] Foehr's group revealed (1) the presence of GPR56 at the leading edge of cell membrane filopodia; (2) colocalized GPR56 and α-actin; and (3) activation of transcription factors, including NF-κB by overexpression of GPR56.[19] Itoh's group recently demonstrated that GPR56 inhibit neural progenitor cell migration by coupling with $G\alpha_{12/13}$ to induce Rho-dependent transcription activation and actin fiber reorganization.[47]

VI. Concluding Remarks

The family of adhesion GPCRs is a newly recognized GPCR subclass. They are predicted to facilitate cell-to-cell or cell-to-matrix interaction. Among the 33 members of human adhesion GPCRs, GPR56 is the first member linked to the devastating human disease BFPP.[2,22] The identification of GPR56 as the BFPP-causing gene raises many unanswered questions. The brain MRIs of BFPP patients often reveal unspecified patchy signal changes in the white matter, suggesting defects in myelination.[22] What is the role of GPR56 in myelination? Furthermore, BFPP resembles cobblestone lissencephaly, a disease often associated with muscular dystrophy. There is no obvious muscle phenotype in individuals harboring mutations in the *GPR56* gene and in *Gpr56* knockout mice.[22,24] Detailed studies are needed to reveal whether there is a subtle abnormality in the skeletal muscles associated with loss of GPR56.

GPR56 is an orphan GPCR. Identification of its ligand will facilitate further study of GPR56 function at cellular level. GPR56 was shown to bind with TG2.[18] The putative ligand of GPR56 is localized in the pial BM in the developing brain.[24] It is yet to be determined whether TG2 is the ligand of GPR56 in the developing brain.

GPR56 has been implicated in tumor cell biology. Evaluation of GPR56 gene expression in human tumors reveals downregulation of GPR56 in some tumor types, while overexpression in others.[18,19,39,44] Analysis of cancer samples in different tumor types and stages also indicates a possible role of GPR56 in tumor progression and metastasis.[18,19,39] GPCRs are one of the most important drug targets for drug therapy. Therefore, it is no doubt that GPR56 will be the future therapeutic targets. Identification of the ligand(s) and elucidation of the protein structure of GPR56 would greatly facilitate therapeutic discovery.

ACKNOWLEDGMENTS

We thank Natalie Strokes for making the figure; Hye Min Yang, Natalie Strokes, and Dr. Lei Xu for reading the manuscript. Z.J. was supported by a Flight Attendant Medical Research Institute Young Clinical Scientist Award. Works in the author's laboratories are in part supported by the NINDS (National Institute of Neurological Disorders and Stroke, K08NS045762 and R01NS057536) and the Charles H. Hood Foundation (Child Health Research Grant, 7/04-6/06).

REFERENCES

1. Liu M, Parker RM, Darby K, Eyre HJ, Copeland NG, Crawford J, et al. GPR56, a novel secretin-like human G-protein-coupled receptor gene. *Genomics* 1999;**55**:296–305.
2. Piao X, Hill RS, Bodell A, Chang BS, Basel-Vanagaite L, Straussberg R, et al. G protein-coupled receptor-dependent development of human frontal cortex. *Science* 2004;**303**:2033–6.
3. Ji TH, Grossmann M, Ji I. G protein-coupled receptors. I. Diversity of receptor-ligand interactions. *J Biol Chem* 1998;**273**:17299–302.
4. Bockaert J, Pin JP. Molecular tinkering of G protein-coupled receptors: an evolutionary success. *EMBO J* 1999;**18**:1723–9.
5. Bjarnadottir TK, Fredriksson R, Hoglund PJ, Gloriam DE, Lagerstrom MC, Schioth HB. The human and mouse repertoire of the adhesion family of G-protein-coupled receptors. *Genomics* 2004;**84**:23–33.
6. Stacey M, Lin HH, Gordon S, McKnight AJ. LNB-TM7, a group of seven-transmembrane proteins related to family-B G-protein-coupled receptors. *Trends Biochem Sci* 2000;**25**:284–9.
7. Fredriksson R, Gloriam DE, Hoglund PJ, Lagerstrom MC, Schioth HB. There exist at least 30 human G-protein-coupled receptors with long Ser/Thr-rich N-termini. *Biochem Biophys Res Commun* 2003;**301**:725–34.

8. Fredriksson R, Lagerstrom MC, Hoglund PJ, Schioth HB. Novel human G protein-coupled receptors with long N-terminals containing GPS domains and Ser/Thr-rich regions. *FEBS Lett* 2002;**531**:407–14.
9. Fredriksson R, Lagerstrom MC, Lundin LG, Schioth HB. The G-protein-coupled receptors in the human genome form five main families. Phylogenetic analysis, paralogon groups, and fingerprints. *Mol Pharmacol* 2003;**63**:1256–72.
10. Krasnoperov V, Lu Y, Buryanovsky L, Neubert TA, Ichtchenko K, Petrenko AG. Post-translational proteolytic processing of the calcium-independent receptor of alpha-latrotoxin (CIRL), a natural chimera of the cell adhesion protein and the G protein-coupled receptor. Role of the G protein-coupled receptor proteolysis site (GPS) motif. *J Biol Chem* 2002;**277**:46518–26.
11. Volynski KE, Silva JP, Lelianova VG, Atiqur Rahman M, Hopkins C, Ushkaryov YA. Latrophilin fragments behave as independent proteins that associate and signal on binding of LTX(N4C). *EMBO J* 2004;**23**:4423–33.
12. Hughes J, Ward CJ, Aspinwall R, Butler R, Harris PC. Identification of a human homologue of the sea urchin receptor for egg jelly: a polycystic kidney disease-like protein. *Hum Mol Genet* 1999;**8**:543–9.
13. Qian F, Boletta A, Bhunia AK, Xu H, Liu L, Ahrabi AK, et al. Cleavage of polycystin-1 requires the receptor for egg jelly domain and is disrupted by human autosomal-dominant polycystic kidney disease 1-associated mutations. *Proc Natl Acad Sci USA* 2002;**99**:16981–6.
14. Abe J, Fukuzawa T, Hirose S. Cleavage of Ig-Hepta at a "SEA" module and at a conserved G protein-coupled receptor proteolytic site. *J Biol Chem* 2002;**277**:23391–8.
15. Jin Z, Tietjen I, Bu L, Liu-Yesucevitz L, Gaur SK, Walsh CA, et al. Disease-associated mutations affect GPR56 protein trafficking and cell surface expression. *Hum Mol Genet* 2007;**16**:1972–85.
16. Lin HH, Chang GW, Davies JQ, Stacey M, Harris J, Gordon S. Autocatalytic cleavage of the EMR2 receptor occurs at a conserved G protein-coupled receptor proteolytic site motif. *J Biol Chem* 2004;**279**:31823–32.
17. Chang GW, Stacey M, Kwakkenbos MJ, Hamann J, Gordon S, Lin HH. Proteolytic cleavage of the EMR2 receptor requires both the extracellular stalk and the GPS motif. *FEBS Lett* 2003;**547**:145–50.
18. Xu L, Begum S, Hearn JD, Hynes RO. GPR56, an atypical G protein-coupled receptor, binds tissue transglutaminase, TG2, and inhibits melanoma tumor growth and metastasis. *Proc Natl Acad Sci USA* 2006;**103**:9023–8.
19. Shashidhar S, Lorente G, Nagavarapu U, Nelson A, Kuo J, Cummins J, et al. GPR56 is a GPCR that is overexpressed in gliomas and functions in tumor cell adhesion. *Oncogene* 2005;**24**:1673–82.
20. Piao X, Basel-Vanagaite L, Straussberg R, Grant PE, Pugh EW, Doheny K, et al. An autosomal recessive form of bilateral frontoparietal polymicrogyria maps to chromosome 16q12.2–21. *Am J Hum Genet* 2002;**70**:1028–33.
21. Chang BS, Piao X, Bodell A, Basel-Vanagaite L, Straussberg R, Dobyns WB, et al. Bilateral frontoparietal polymicrogyria: clinical and radiological features in 10 families with linkage to chromosome 16. *Ann Neurol* 2003;**53**:596–606.
22. Piao X, Chang BS, Bodell A, Woods K, Benzeev B, Topcu M, et al. Genotype–phenotype analysis of human frontoparietal polymicrogyria syndromes. *Ann Neurol* 2005;**58**:680–7.
23. Parrini E, Ferrari AR, Dorn T, Walsh CA, Guerrini R. Bilateral frontoparietal polymicrogyria, Lennox–Gastaut syndrome, and GPR56 gene mutations. *Epilepsia* 2008;. doi:10.1111/j.1528-1167.2008.01787.x.
24. Li S, Jin Z, Koirala S, Bu L, Xu L, Hynes RO, et al. GPR56 regulates pial basement membrane integrity and cortical lamination. *J Neurosci* 2008;**28**:5817–26.

25. Olson EC, Walsh CA. Smooth, rough and upside-down neocortical development. *Curr Opin Genet Dev* 2002;**12**:320–7.
26. Kobayashi K, Nakahori Y, Miyake M, Matsumura K, Kondo-Iida E, Nomura Y, et al. An ancient retrotransposal insertion causes Fukuyama-type congenital muscular dystrophy. *Nature* 1998;**394**:388–92.
27. Yoshida A, Kobayashi K, Manya H, Taniguchi K, Kano H, Mizuno M, et al. Muscular dystrophy and neuronal migration disorder caused by mutations in a glycosyltransferase, POMGnT1. *Dev Cell* 2001;**1**:717–24.
28. Michele DE, Barresi R, Kanagawa M, Saito F, Cohn RD, Satz JS, et al. Post-translational disruption of dystroglycan–ligand interactions in congenital muscular dystrophies. *Nature* 2002;**418**:417–22.
29. Georges-Labouesse E, Mark M, Messaddeq N, Gansmuller A. Essential role of alpha 6 integrins in cortical and retinal lamination. *Curr Biol* 1998;**8**:983–6.
30. De Arcangelis A, Mark M, Kreidberg J, Sorokin L, Georges-Labouesse E. Synergistic activities of alpha3 and alpha6 integrins are required during apical ectodermal ridge formation and organogenesis in the mouse. *Development* 1999;**126**:3957–68.
31. Graus-Porta D, Blaess S, Senften M, Littlewood-Evans A, Damsky C, Huang Z, et al. Beta1-class integrins regulate the development of laminae and folia in the cerebral and cerebellar cortex. *Neuron* 2001;**31**:367–79.
32. Beggs HE, Schahin-Reed D, Zang K, Goebbels S, Nave KA, Gorski J, et al. FAK deficiency in cells contributing to the basal lamina results in cortical abnormalities resembling congenital muscular dystrophies. *Neuron* 2003;**40**:501–14.
33. Niewmierzycka A, Mills J, St-Arnaud R, Dedhar S, Reichardt LF. Integrin-linked kinase deletion from mouse cortex results in cortical lamination defects resembling cobblestone lissencephaly. *J Neurosci* 2005;**25**:7022–31.
34. Costell M, Gustafsson E, Aszodi A, Morgelin M, Bloch W, Hunziker E, et al. Perlecan maintains the integrity of cartilage and some basement membranes. *J Cell Biol* 1999;**147**:1109–22.
35. Sievers J, Pehlemann FW, Gude S, Berry M. Meningeal cells organize the superficial glia limitans of the cerebellum and produce components of both the interstitial matrix and the basement membrane. *J Neurocytol* 1994;**23**:135–49.
36. Halfter W, Dong S, Yip YP, Willem M, Mayer U. A critical function of the pial basement membrane in cortical histogenesis. *J Neurosci* 2002;**22**:6029–40.
37. Haubst N, Georges-Labouesse E, De Arcangelis A, Mayer U, Gotz M. Basement membrane attachment is dispensable for radial glial cell fate and for proliferation, but affects positioning of neuronal subtypes. *Development* 2006;**133**:3245–54.
38. Hu H, Yang Y, Eade A, Xiong Y, Qi Y. Breaches of the pial basement membrane and disappearance of the glia limitans during development underlie the cortical lamination defect in the mouse model of muscle–eye–brain disease. *J Comp Neurol* 2007;**501**:168–83.
39. Zendman AJ, Cornelissen IM, Weidle UH, Ruiter DJ, van Muijen GN. TM7XN1, a novel human EGF-TM7-like cDNA, detected with mRNA differential display using human melanoma cell lines with different metastatic potential. *FEBS Lett* 1999;**446**:292–8.
40. Hand D, Elliott BM, Griffin M. Correlation of changes in transglutaminase activity and polyamine content of neoplastic tissue during the metastatic process. *Biochim Biophys Acta* 1987;**930**:432–7.
41. Knight CR, Rees RC, Elliott BM, Griffin M. The existence of an inactive form of transglutaminase within metastasising tumours. *Biochim Biophys Acta* 1990;**1053**:13–20.
42. Birckbichler PJ, Bonner RB, Hurst RE, Bane BL, Pitha JV, Hemstreet III GP. Loss of tissue transglutaminase as a biomarker for prostate adenocarcinoma. *Cancer* 2000;**89**:412–23.

43. Barnes RN, Bungay PJ, Elliott BM, Walton PL, Griffin M. Alterations in the distribution and activity of transglutaminase during tumour growth and metastasis. *Carcinogenesis* 1985;**6**:459–63.
44. Ke N, Sundaram R, Liu G, Chionis J, Fan W, Rogers C, et al. Orphan G protein-coupled receptor GPR56 plays a role in cell transformation and tumorigenesis involving the cell adhesion pathway. *Mol Cancer Ther* 2007;**6**:1840–50.
45. Little KD, Hemler ME, Stipp CS. Dynamic regulation of a GPCR–tetraspanin–G protein complex on intact cells: central role of CD81 in facilitating GPR56-Galpha q/11 association. *Mol Biol Cell* 2004;**15**:2375–87.
46. Levy S, Shoham T. The tetraspanin web modulates immune-signalling complexes. *Nat Rev Immunol* 2005;**5**:136–48.
47. Iguchi T, Sakata K, Yoshizaki K, Tago K, Mizuno N, Itoh H. Orphan G protein-coupled receptor GPR56 regulates neural progenitor cell migration via a G alpha 12/13 and Rho pathway. *J Biol Chem* 2008;**283**:14469–78.

V2R Mutations and Nephrogenic Diabetes Insipidus

DANIEL G. BICHET[*,†,‡]

[*]*Canada Research Chair in Genetics of Renal Diseases, Groupe d'Étude des Protéines Membranaires, Montréal, Québec, Canada H4J 1C5*

[†]*Department of Medicine and Physiology, Université de Montréal, Research Center, Montréal, Québec, Canada H4J 1C5*

[‡]*Nephrology Service, Hôpital du Sacré-Coeur de Montréal, Montréal, Québec, Canada H4J 1C5*

I. Cellular Actions of Vasopressin 16
II. Rareness and Diversity of *AVPR2* Mutations 20
III. Most Mutant V2 Receptors Are Not Transported to the Cell Membrane and Are Retained in the Intracellular Compartments 22
IV. Nonpeptide Vasopressin Receptor Antagonists Act as Pharmacological Chaperones to Functionally Rescue Misfolded Mutant V2 Receptors Responsible for X-Linked NDI 24
V. Gain of Function of the Vasopressin V2 Receptor: Nephrogenic Syndrome of Inappropriate Antidiuresis 24
References 26

Nephrogenic diabetes insipidus (NDI), which can be inherited or acquired, is characterized by an inability to concentrate urine despite normal or elevated plasma concentrations of the antidiuretic hormone, arginine vasopressin (AVP). Polyuria, with hyposthenuria, and polydipsia are the cardinal clinical manifestations of the disease. Nephrogenic failure to concentrate urine maximally may be due to a defect in vasopressin-induced water permeability of the distal tubules and collecting ducts, to insufficient buildup of the corticopapillary interstitial osmotic gradient, or to a combination of these two factors. Thus, the broadest definition of the term NDI embraces any antidiuretic hormone-resistant urinary-concentrating defect, including medullary disease with low interstitial osmolality, renal failure, and osmotic diuresis. About 90% of patients with congenital NDI are males with X-linked recessive NDI (OMIM 304800)[1] and have mutations in the AVP receptor 2 (*AVPR2*) gene that codes for the vasopressin V_2 receptor; the gene is located in chromosome region Xq28. In about 10% of the families studied, congenital NDI has an autosomal recessive or autosomal dominant mode of inheritance (OMIM 222000 and 125800)[1]. Mutations have been identified in the aquaporin-2 gene (*AQP2*, OMIM 107777)[1], which is located in chromosome region 12q13 and codes for the vasopressin-sensitive water channel.

NDI is clinically distinguishable from neurohypophyseal diabetes insipidus (OMIM 125700[1]; also referred to as central or neurogenic diabetes insipidus) by a lack of response to exogenous AVP and by plasma levels of AVP that rise normally with increase in plasma osmolality. Hereditary neurohypophyseal diabetes insipidus is secondary to mutations in the gene encoding AVP (OMIM 192340)[1]. Neurohypophyseal diabetes insipidus is also a component of autosomal recessive Wolfram syndrome 1 or DIDMOAD syndrome (*d*iabetes *i*nsipidus, *d*iabetes *m*ellitus, *o*ptic *a*trophy, and *d*eafness) (OMIM 222300)[1], an autosomal recessive disorder. Other inherited disorders with complex polyuro-polydipsic syndrome with loss of water, sodium, chloride, calcium, magnesium, and potassium include Bartter syndrome (OMIM 601678)[1] and cystinosis (OMIM 219800)[1], while long-term lithium administration is the main cause of acquired NDI. Here, we use the gene symbols approved by the HUGO Gene Nomenclature Committee (http://www.gene.ucl.ac.uk/nomenclature) and provide OMIM entry numbers [OMIM (Online Mendelian Inheritance in Man)[1]; McKusick-Nathans Institute for Genetic Medicine, Johns Hopkins University (Baltimore, MD) and National Center for Biotechnology Information, National Library of Medicine (Bethesda, MD), 2000; World Wide Web URL: http://www.ncbi.nlm.nih.gov/omim/].

I. Cellular Actions of Vasopressin

Homologues of vasopressin and oxytocin have evolved over 700 million years and have been identified in insects to vertebrates.[2–3] The *cis* and *trans* components important for vasopressin and oxytocin expression in magnocellular neurons have been conserved over 450 million years in the pufferfish isotocin and rat oxytocin genes.[4,5] Among these distant taxa (hydra, worms, insects, and vertebrates), oxytocin- and vasopressin-related peptides also play a general role in the modulation of social and reproductive behavior.[2] In contrast to this apparent conservation of function, the specific behaviors affected by these neuropeptides are notably species specific.[2] The neurohypophyseal hormone arginine vasopressin (AVP) has multiple actions, including the inhibition of diuresis, contraction of smooth muscle, aggregation of platelets, stimulation of liver glycogenolysis, modulation of adrenocorticotropic hormone release from the pituitary, and central regulation of somatic functions (thermoregulation and blood pressure) and modulation of social and reproductive behavior. These multiple actions of AVP can be explained by the interaction of AVP with at least three types of G-protein-coupled receptors: the V_{1a} (vascular, hepatic, and brain) and V_{1b} (anterior pituitary) receptors act through phosphatidylinositol hydrolysis to mobilize calcium, and the V_2 (kidney) receptor is coupled to adenylate cyclase.[6–8]

The transfer of water across the principal cells of the collecting ducts is now known at such a detailed level that billions of molecules of water traversing the membrane can be represented; see useful teaching tools at http://www.mpibpc.gwdg.de/abteilungen/073/gallery.html and http://www.ks.uiuc.edu/research/aquaporins. The 2003 Nobel Prize in chemistry was awarded to Peter Agre and Roderick MacKinnon, who solved two complementary problems presented by the cell membrane: How does a cell let one type of ion through the lipid membrane to the exclusion of other ions? And how does it permeate water without ions? This contributed to a momentum and renewed interest in basic discoveries related to the transport of water and indirectly to diabetes insipidus. The first step in the action of AVP on water excretion is its binding to AVP type 2 receptors (AVPR2) (hereafter referred to as V_2 receptors) on the basolateral membrane of the collecting duct cells (Fig. 1). The human *AVPR2* gene that codes for the V_2 receptor is located in chromosome region Xq28 and has three exons and two small introns.[9,10] The sequence of the cDNA predicts a polypeptide of 371 amino acids with seven transmembrane, four extracellular, and four cytoplasmic domains (Fig. 2). The V_2 receptor is one of 701 members of the rhodopsin family within the superfamily of guanine-nucleotide (G)-protein-coupled receptors[11] (see also the perspective by Perez[12] and recent comments on X-ray structure breakthroughs in the transmembrane-spanning region[13]). The activation of the V_2 receptor on renal collecting tubules stimulates adenylyl cyclase via the stimulatory G protein (Gs) and promotes the cyclic adenosine monophosphate (cAMP)-mediated incorporation of water channels into the luminal surface of these cells. There are two ubiquitously expressed intracellular cAMP receptors: (i) the classical protein kinase A (PKA) that is a cAMP-dependent protein kinase and (ii) the recently discovered exchange protein directly activated by cAMP that is a cAMP-regulated guanine nucleotide exchange factor. Both of these receptors contain an evolutionarily conserved cAMP-binding domain that acts as a molecular switch for sensing intracellular cAMP levels to control diverse biological functions.[14] Several proteins participating in the control of cAMP-dependent AQP2 trafficking have been identified; for example, A-kinase anchoring proteins tethering PKA to cellular compartments; phosphodiesterases regulating the local cAMP level; cytoskeletal components such as F-actin and microtubules; small GTPases of the Rho family controlling cytoskeletal dynamics; motor proteins transporting AQP2-bearing vesicles to and from the plasma membrane

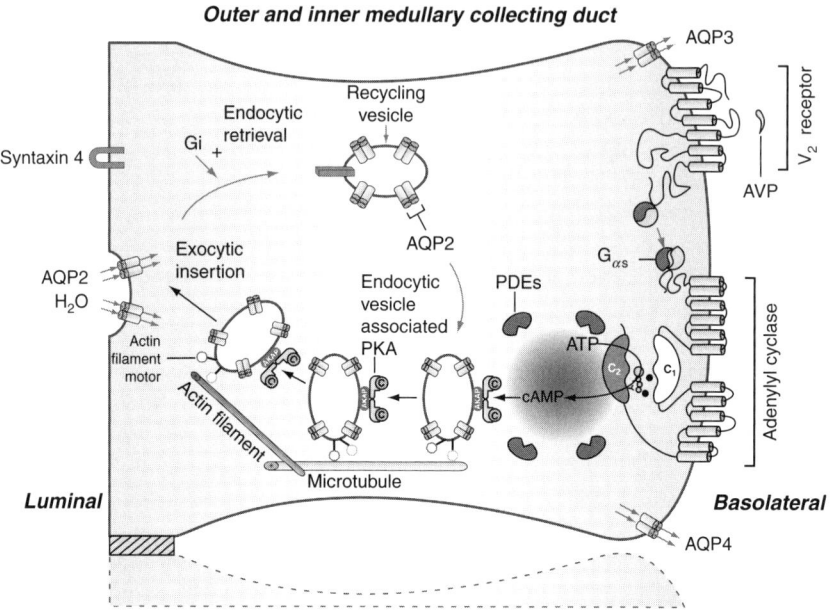

Fig. 1. Schematic representation of the effect of vasopressin (AVP) to increase water permeability in the principal cells of the collecting duct. AVP is bound to the V_2 receptor (a G-protein-linked receptor) on the basolateral membrane. The basic process of G-protein-coupled receptor signaling consists of three steps: a hepta-helical receptor that detects a ligand (in this case, AVP) in the extracellular milieu, a G protein ($G_{\alpha s}$) that dissociates into α subunits bound to GTP and $\beta \gamma$ subunits after interaction with the ligand-bound receptor, and an effector (in this case, adenylyl cyclase) that interacts with dissociated G-protein subunits to generate small-molecule second messengers. AVP activates adenylyl cyclase, increasing the intracellular concentration of cAMP. The topology of adenylyl cyclase is characterized by two tandem repeats of six hydrophobic transmembrane domains separated by a large cytoplasmic loop and terminates in a large intracellular tail. The dimeric structure (C_1 and C_2) of the catalytic domains is represented. Conversion of ATP to cAMP takes place at the dimer interface. Two aspartate residues (in C_1) coordinate two metal cofactors (Mg^{2+} or Mn^{2+} represented here as two small black circles), which enable the catalytic function of the enzyme.[65] Adenosine is shown as an open circle and the three phosphate groups (ATP) are shown as smaller open circles. Protein kinase A (PKA) is the target of the generated cAMP. The binding of cAMP to the regulatory subunits of PKA induces a conformational change, causing these subunits to dissociate from the catalytic subunits. These activated subunits (C) as shown here are anchored to an aquaporin-2 (AQP2)-containing endocytic vesicle via an A-kinase anchoring protein. The local concentration and distribution of the cAMP gradient is limited by phosphodiesterases (PDEs). Cytoplasmic vesicles carrying the water channels (represented as homotetrameric complexes) are fused to the luminal membrane in response to AVP, thereby increasing the water permeability of this membrane. The dissociation of the A-kinase anchoring protein from the endocytic vesicle is not represented. Microtubules and actin filaments are necessary for vesicle movement toward the membrane. When AVP is not available, AQP2 water channels are retrieved by an endocytic process, and water permeability returns to its original low rate. Aquaporin-3 (AQP3) and aquaporin-4 (AQP4) water channels are expressed constitutively at the basolateral membrane.

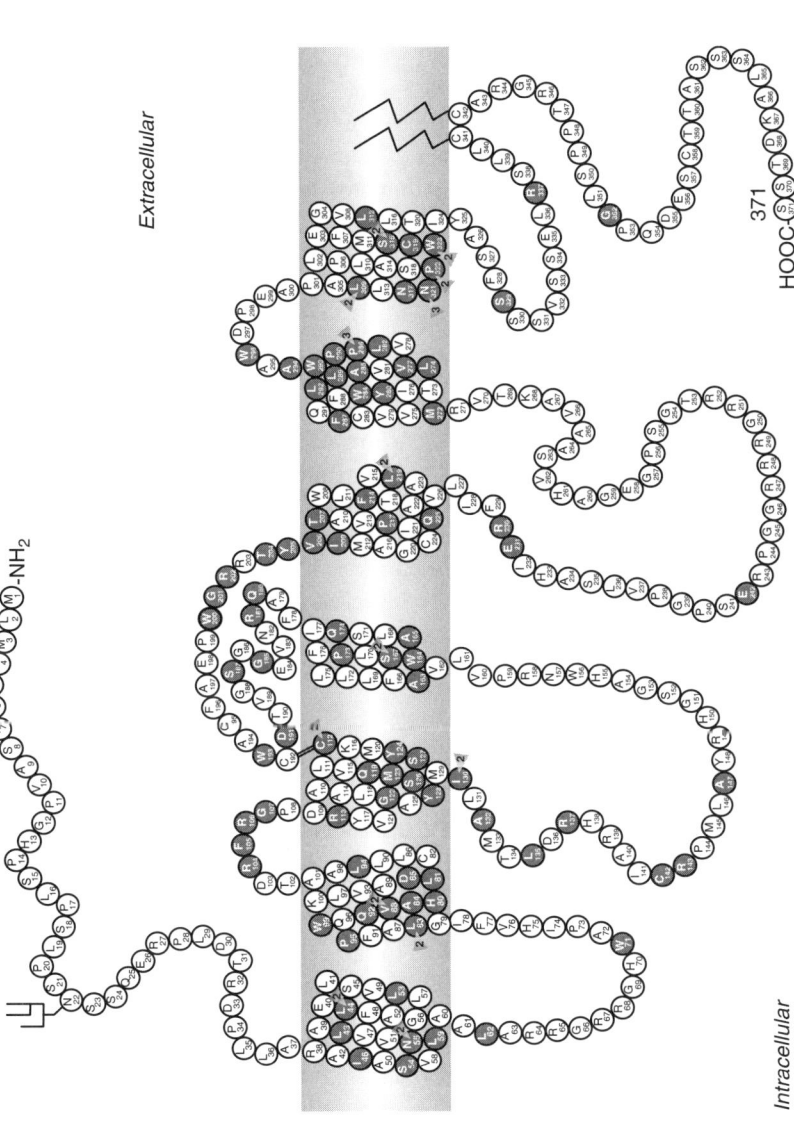

FIG. 2. Schematic representation of the V_2 receptor and identification of 193 putative disease-causing *AVPR2* mutations. Predicted amino acids are shown as the one-letter amino acid code. A solid symbol indicates a codon with a missense or nonsense mutation; a number indicates more than one mutation in the same codon; other types of mutations are not indicated in the figure. The extracellular, transmembrane, and cytoplasmic domains are defined according to Mouillac et al.[66] There are 95 missense, 18 nonsense, 46 frameshift deletion or insertion, 7 in-frame deletion or insertion, 4 splice-site, and 22 large deletion mutations, and 1 complex mutation. (See Color Insert.)

for exocytic insertion and endocytic retrieval; SNAREs inducing membrane fusions, hsc70, a chaperone important for endocytic retrieval. These processes are the molecular basis of the vasopressin-induced increase in the osmotic water permeability of the apical membrane of the collecting tubule.[15–17]

AVP also increases the water reabsorptive capacity of the kidney by regulating the urea transporter UT-A1 that is present in the inner medullary collecting duct, predominantly in its terminal part.[18,19] AVP also increases the permeability of principal collecting duct cells to sodium.[20]

In summary, in the absence of AVP stimulation, collecting duct epithelia exhibit very low permeabilities to sodium urea and water. These specialized permeability properties permit the excretion of large volumes of hypotonic urine formed during intervals of water diuresis. By contrast, AVP stimulation of the principal cells of the collecting ducts leads to selective increases in the permeability of the apical membrane to water, urea, and sodium.

These actions of vasopressin in the distal nephron are possibly modulated by prostaglandin E2, nitric oxide,[21] and by luminal calcium concentration. High levels of E-prostanoid-3 receptors are expressed in the kidney.[22] However, mice lacking E-prostanoid-3 receptors for prostaglandin E2 were found to have quasi-normal regulation of urine volume and osmolality in response to various physiological stimuli.[22] An apical calcium/polycation receptor protein expressed in the terminal portion of the inner medullary collecting duct of the rat has been shown to reduce AVP-elicited osmotic water permeability when luminal calcium concentration rises.[23] This possible link between calcium and water metabolism may play a role in the pathogenesis of renal stone formation.[23]

II. Rareness and Diversity of *AVPR2* Mutations

X-linked nephrogenic diabetes insipidus (NDI) is generally a rare disease in which the affected male patients do not concentrate their urine after administration of AVP.[24] Because this form is a rare, recessive X-linked disease, females are unlikely to be affected, but heterozygous females can exhibit variable degrees of polyuria and polydipsia because of skewed X-chromosome inactivation. In Quebec, the incidence of this disease among males was estimated to be approximately 8.8 in 1,000,000 male live births.[25] A founder effect of two particular *AVPR2* mutations,[26] one in Ulster Scot immigrants (the Hopewell mutation, W71X) and one in a large Utah kindred (the Cannon pedigree) results in an elevated prevalence of X-linked NDI in their descendants in certain communities in Nova Scotia, Canada and in Utah, USA.[26]

These founder mutations have now spread all over the North American continent. To date, we have identified the W71X mutation in 42 affected males who reside predominantly in the Maritime Provinces of Nova Scotia and New Brunswick, and the L312X mutation has been identified in eight affected males who reside in the central USA. We know of 98 living affected males of the Hopewell kindred and 18 living affected males of the Cannon pedigree. We also determined that the historical case report by Perry et al.[27] was related to the Hopewell pedigree and had the W71X mutation (Fig. 3).

To date, about 200 putative disease-causing *AVPR2* mutations have been published in over 300 NDI families (Fig. 2).[28,29] Approximately, half of the mutations are missense mutations. Frameshift mutations owing to nucleotide deletions or insertions (24%), nonsense mutations (9%), large deletions (11%), in-frame deletions or insertions (4%), splice-site mutations (2%), and one complex mutation account for the remainder of the mutations.[29–31] Mutations have been identified in every domain, but on a per nucleotide basis, about twice

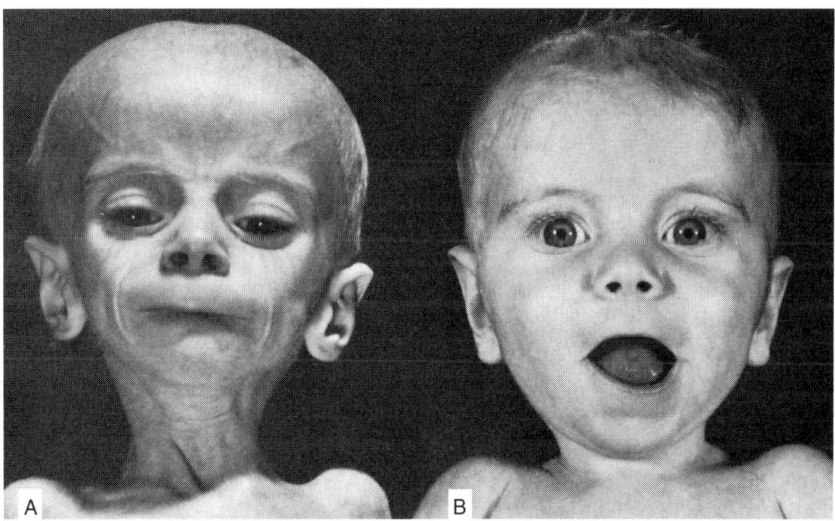

FIG. 3. A typical historical picture of a dehydrated and malnourished infant with nephrogenic diabetes insipidus (A), looking healthy after rehydration and improved nutrition (B). This infant died a few years later due to repeated episodes of dehydration. This report was published years before the identification of the *AVPR2* gene. We were contacted by the mother and sister of this patient and we were able to reconstruct and link this family to the large Hopewell kindred[26] (Bichet and Arthus, unpublished data). This photograph is Fig. 2 of Perry et al.[27] reproduced with permission from the *New England Journal of Medicine*.

as many mutations occur in transmembrane domains compared with the extracellular or intracellular domains. We previously identified private mutations, recurrent mutations, and mechanisms of mutagenesis.[25,32] Ten recurrent mutations (D85N, V88M, R113W, Y128S, R137H, S167L, R181C, R202C, A294P, and S315R) were found in 35 ancestrally independent families. The occurrence of the same mutation on different haplotypes was considered evidence for recurrent mutation. In addition, the most frequent mutations (D85N, V88N, R113W, R137H, S167L, R181C, and R202C) occurred at potential mutational hot spots (a C-to-T or G-to-A nucleotide substitution at a CpG dinucleotide).

III. Most Mutant V2 Receptors Are Not Transported to the Cell Membrane and Are Retained in the Intracellular Compartments

Classification of the defects of naturally occurring mutant human V_2 receptors can be based on a similar scheme to that used for the low-density lipoprotein receptor. Mutations have been grouped according to the function and subcellular localization of the mutant protein whose cDNA has been transiently transfected in a heterologous expression system.[33] Using this classification, type 1 mutant V_2 receptors reach the cell surface but display impaired ligand binding and are consequently unable to induce normal cAMP production. The dose-dependent AVP-stimulated cAMP production could be compared with the cAMP production obtained with the wild-type receptor using a protein-based bioluminescence resonance energy transfer cAMP biosensor that allows the measurement of cAMP in living cells.[34] The presence of mutant V_2 receptors on the surface of transfected cells can be determined pharmacologically. By carrying out saturation binding experiments using tritiated AVP, the number of cell surface mutant V_2 receptors and their apparent binding affinity can be compared with that of the wild-type receptor. In addition, the presence of cell surface receptors can be assessed directly by using immunodetection strategies to visualize epitope-tagged receptors in whole-cell immunofluorescence assays.

Type 2 mutant receptors have defective intracellular transport. This phenotype is confirmed by carrying out, in parallel, immunofluorescence experiments on cells that are intact (to demonstrate the absence of cell surface receptors) or permeabilized (to confirm the presence of intracellular receptor pools). In addition, protein expression is confirmed by Western blot analysis of membrane preparations from transfected cells. It is likely that these mutant type 2 receptors accumulate in a pre-Golgi compartment because they are initially glycosylated but fail to undergo glycosyl-trimming maturation.

Type 3 mutant receptors are ineffectively transcribed and lead to unstable mRNAs which are rapidly degraded. This subgroup seems to be rare since Northern blot analysis of cells expressing mutant *AVPR2* receptors showed mRNAs of normal quantity and molecular size.

Most of the *AVPR2* mutants that we and other investigators have tested are type 2 mutant receptors. They did not reach the cell membrane and were trapped in the interior of the cell.[35–39] Other mutant G-protein-coupled receptors[40] and gene products causing genetic disorders are also characterized by protein misfolding. Mutations that affect the folding of secretory proteins, integral plasma membrane proteins, or enzymes destined to the endoplasmic reticulum, Golgi complex, and lysosomes result in loss-of-function phenotypes irrespective of their direct impact on protein function because these mutant proteins are prevented from reaching their final destination.[41,42] Folding in the endoplasmic reticulum is the limiting step: mutant proteins which fail to correctly fold are initially retained in the endoplasmic reticulum and subsequently often degraded. Key proteins involved in the urine countercurrent mechanisms are good examples of this basic mechanism of misfolding. AQP2 mutations responsible for autosomal recessive NDI are characterized by misrouting of the misfolded mutant proteins and are trapped in the endoplasmic reticulum.[43] Other mutant renal membrane proteins that are responsible for Gitelman syndrome,[44] Bartter syndrome,[45,46] and cystinuria[47] are also retained in the endoplasmic reticulum.

The *AVPR2* missense mutations are likely to impair folding and to lead to rapid degradation of the misfolded polypeptide and not to the accumulation of toxic aggregates (as is the case for AVP mutants that cause neurohypophyseal diabetes insipidus), because the other important functions of the principal cells of the collecting duct (where *AVPR2* is expressed) are entirely normal. These cells express the epithelial sodium channel (ENac). Decreased function of this channel results in a sodium-losing state.[48] This has not been observed in patients with *AVPR2* mutations. However, recent data showed that dDAVP could not stimulate sodium reabsorption in male patients with NDI bearing *AVPR2* mutations[20] but this is a V2R-specific effect.[49] By contrast, another type of conformational disease is characterized by the toxic retention of the misfolded protein. The relatively common Z mutation in α1-antitrypsin deficiency not only causes retention of the mutant protein in the endoplasmic reticulum but also affects the secondary structure by insertion of the reactive center loop of one molecule into a destabilized β sheet of a second molecule.[50] These polymers clog up the endoplasmic reticulum of hepatocytes and lead to cell death and juvenile hepatitis, cirrhosis, and hepatocarcinomas in these patients.[51]

IV. Nonpeptide Vasopressin Receptor Antagonists Act as Pharmacological Chaperones to Functionally Rescue Misfolded Mutant V2 Receptors Responsible for X-Linked NDI

If the misfolded protein/traffic problem responsible for so many human genetic diseases can be overcome and the mutant protein can be transported out of the endoplasmic reticulum to its final destination, these mutant proteins might be sufficiently functional.[52] Therefore, using pharmacological chaperones to promote escape from the endoplasmic reticulum is a possible therapeutic approach.[36,39,41,53] We used selective nonpeptide V_2 and V_1 receptor antagonists to rescue the cell surface expression and function of naturally occurring misfolded human V_2 receptors.[35] Since the beneficial effect of nonpeptide V_2 antagonists could be secondary to prevention or interference with endocytosis, we studied the R137H mutant previously reported to lead to constitutive endocytosis.[54] We found that the antagonist did not prevent the constitutive β-arrestin-mediated endocytosis.[36] These results indicate that as for other *AVPR2* mutants, the beneficial effects of the treatment result from the action of the pharmacological chaperones. In clinical studies, we administered a nonpeptide vasopressin antagonist SR49059 to five adult NDI patients bearing the del62–64, R137H, or W164S mutation. SR49059 significantly decreased urine volume and water intake and increased urine osmolality while sodium, potassium, and creatinine excretions and plasma sodium were constant throughout the study[55] (Fig. 4). This new therapeutic approach could be applied to the treatment of several hereditary diseases resulting from errors in protein folding and kinesis.[52,53]

Since most human gene therapy experiments using viruses to deliver and integrate DNA into host cells are potentially dangerous,[56] other treatments are being actively pursued. Schoneberg and colleagues[57] used aminoglycoside antibiotics because of their ability to suppress premature termination codons.[58] They demonstrated that geneticin, a potent aminoglycoside antibiotic, increased AVP-stimulated cAMP in cultured collecting duct cells prepared from E242X mutant mice. The urine-concentrating ability of heterozygous mutant mice was also improved.

V. Gain of Function of the Vasopressin V2 Receptor: Nephrogenic Syndrome of Inappropriate Antidiuresis

The clinical phenotype here is opposite to NDI. Rare cases of infants or adults with hyponatremia, concentrated urine, and suppressed AVP plasma concentrations have been described bearing the mutations R137C or R137L in

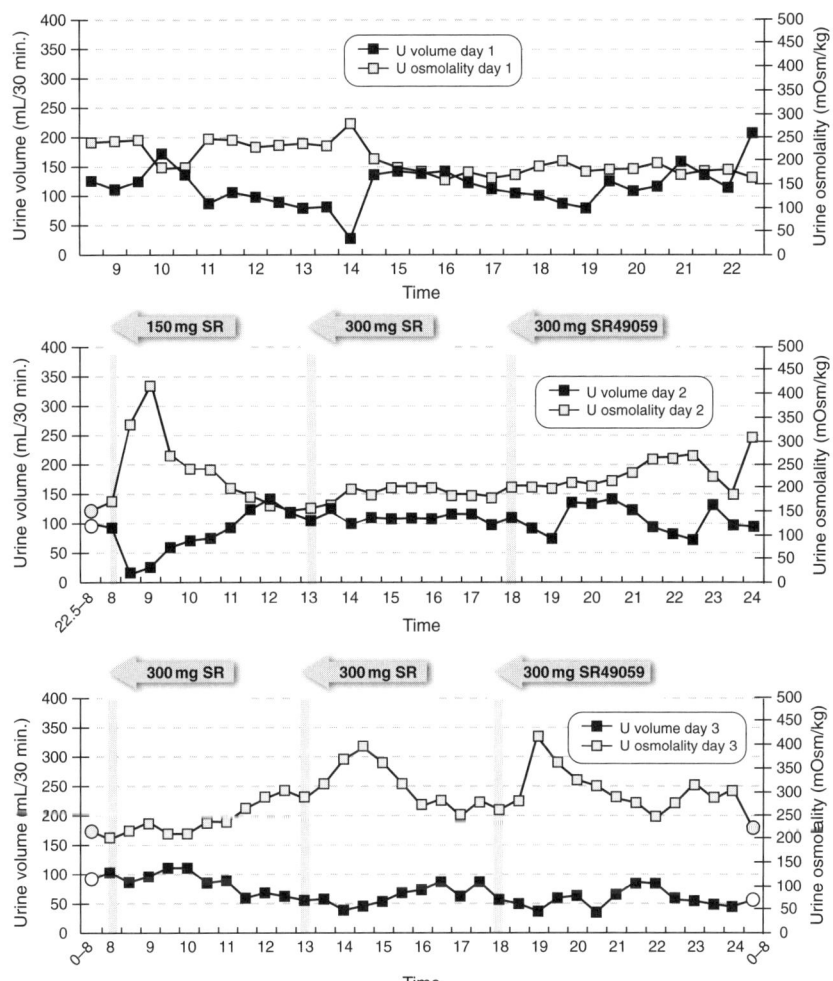

FIG. 4. Urine volume and osmolality before (day 1) and after (days 2 and 3) SR49059 administration to a patient bearing the R137H mutation. Note the mirror image of urine volume and osmolality observed on days 2 and 3. Urine volume and osmolalities that were obtained during the control, second, and third nights are indicated by round circles. These data were obtained from 9:30 p.m. to 8:00 a.m. for the patient described here. Data from Bernier et al.[55] with permission from the *Journal of the American Society of Nephrology*.

their *AVPR2* gene.[59–62] It is interesting to note that another mutation in the same codon (R137H) is a relatively frequent mutation causing classical NDI, albeit the phenotype may be milder in some patients.[63] With cell-based assays, both R137C and R137L were found to have elevated basal signaling through the cAMP pathway and to interact with β-arrestins in an agonist-independent manner.[64] It is my opinion that *AVPR2* mutations with gain of function are extremely rare. We have sequenced the *AVPR2* gene in many patients with hyponatremia and never found a mutation. By contrast, we continue to identify new and recurrent loss-of-function *AVPR2* mutations in patients with classical NDI.

REFERENCES

1. OMIM (Online Mendelian Inheritance in Man). McKusick-Nathans Institute for Genetic Medicine, Johns Hopkins University (Baltimore, MD) and National Center for Biotechnology Information, National Library of Medicine (Bethesda, MD), 2000. World Wide Web URL: http://www.ncbi.nlm.nih.gov/omim/; 2000.
2. Donaldson ZR, Young LJ. Oxytocin, vasopressin, and the neurogenetics of sociality. *Science* 2008;**322**:900–4.
3. Gwee PC, Amemiya CT, Brenner S, Venkatesh B. Sequence and organization of coelacanth neurohypophysial hormone genes: evolutionary history of the vertebrate neurohypophysial hormone gene locus. *BMC Evol Biol* 2008;**8**:93.
4. Venkatesh B, Si-Hoe SL, Murphy D, Brenner S. Transgenic rats reveal functional conservation of regulatory controls between the Fugu isotocin and rat oxytocin genes. *Proc Natl Acad Sci USA* 1997;**94**:12462–6.
5. Tessmar-Raible K, Raible F, Christodoulou F, Guy K, Rembold M, Hausen H, et al. Conserved sensory-neurosecretory cell types in annelid and fish forebrain: insights into hypothalamus evolution. *Cell* 2007;**129**:1389–400.
6. Thibonnier M, Coles P, Thibonnier A, Shoham M. The basic and clinical pharmacology of nonpeptide vasopressin receptor antagonists. *Annu Rev Pharmacol Toxicol* 2001;**41**:175–202.
7. Serradeil-Le Gal C, Wagnon J, Valette G, Garcia G, Pascal M, Maffrand JP, et al. Nonpeptide vasopressin receptor antagonists: development of selective and orally active V1a, V2 and V1b receptor ligands. *Prog Brain Res* 2002;**139**:197–210.
8. Walum H, Westberg L, Henningsson S, Neiderhiser JM, Reiss D, Igl W, et al. Genetic variation in the vasopressin receptor 1a gene (AVPR1A) associates with pair-bonding behavior in humans. *Proc Natl Acad Sci USA* 2008;**105**:14153–6.
9. Birnbaumer M, Seibold A, Gilbert S, Ishido M, Barberis C, Antaramian A, et al. Molecular cloning of the receptor for human antidiuretic hormone. *Nature* 1992;**357**:333–5.
10. Seibold A, Brabet P, Rosenthal W, Birnbaumer M. Structure and chromosomal localization of the human antidiuretic hormone receptor gene. *Am J Hum Genet* 1992;**51**:1078–83.
11. Fredriksson R, Lagerstrom MC, Lundin LG, Schioth HB. The G-protein-coupled receptors in the human genome form five main families. Phylogenetic analysis, paralogon groups, and fingerprints. *Mol Pharmacol* 2003;**63**:1256–72.
12. Perez DM. The evolutionarily triumphant G-protein-coupled receptor. *Mol Pharmacol* 2003;**63**:1202–5.
13. Topiol S, Sabio M. X-ray structure breakthroughs in the GPCR transmembrane region. *Biochem Pharmacol* 2009;**78**:11–20.
14. Rehmann H, Wittinghofer A, Bos JL. Capturing cyclic nucleotides in action: snapshots from crystallographic studies. *Nat Rev Mol Cell Biol* 2007;**8**:63–73.
15. Nielsen S, Frokiaer J, Marples D, Kwon TH, Agre P, Knepper MA. Aquaporins in the kidney: from molecules to medicine. *Physiol Rev* 2002;**82**:205–44.

16. Boone M, Deen PM. Physiology and pathophysiology of the vasopressin-regulated renal water reabsorption. *Pflugers Arch* 2008;**456**:1005–24.
17. Nedvetsky PI, Tamma G, Beulshausen S, Valenti G, Rosenthal W, Klussmann E. Regulation of aquaporin-2 trafficking. *Handb Exp Pharmacol* 2009;133–57.
18. Yang B, Bankir L, Gillespie A, Epstein CJ, Verkman AS. Urea-selective concentrating defect in transgenic mice lacking urea transporter UT-B. *J Biol Chem* 2002;**277**:10633–7.
19. Smith CP. Mammalian urea transporters. *Exp Physiol* 2009;**94**:180–5.
20. Bankir L, Fernandes S, Bardoux P, Bouby N, Bichet DG. Vasopressin-V2 receptor stimulation reduces sodium excretion in healthy humans. *J Am Soc Nephrol* 2005;**16**:1920–8.
21. Morishita T, Tsutsui M, Shimokawa H, Sabanai K, Tasaki H, Suda O, et al. Nephrogenic diabetes insipidus in mice lacking all nitric oxide synthase isoforms. *Proc Natl Acad Sci USA* 2005;**102**:10616–21.
22. Fleming EF, Athirakul K, Oliverio MI, Key M, Goulet J, Koller BH, et al. Urinary concentrating function in mice lacking EP3 receptors for prostaglandin E2. *Am J Physiol* 1998;**275**:F955–61.
23. Sands JM, Naruse M, Baum M, Jo I, Hebert SC, Brown EM, et al. Apical extracellular calcium/polyvalent cation-sensing receptor regulates vasopressin-elicited water permeability in rat kidney inner medullary collecting duct. *J Clin Invest* 1997;**99**:1399–405.
24. Bichet DG, Fujiwara TM. Nephrogenic diabetes insipidus. In: Scriver CR, Beaudet AL, Sly D, Vallee D, Childs B, Kinzler KW, Vogelstein B, editors. *The metabolic and molecular bases of inherited disease*. New York: McGraw-Hill; 2001. p. 4181–204.
25. Arthus M-F, Lonergan M, Crumley MJ, Naumova AK, Morin D, De Marco L, et al. Report of 33 novel *AVPR2* mutations and analysis of 117 families with X-linked nephrogenic diabetes insipidus. *J Am Soc Nephrol* 2000;**11**:1044–54.
26. Bichet DG, Arthus M-F, Lonergan M, Hendy GN, Paradis AJ, Fujiwara T.M, et al. X-linked nephrogenic diabetes insipidus mutations in North America and the Hopewell hypothesis. *J Clin Invest* 1993;**92**:1262–8.
27. Perry TL, Robinson GC, Teasdale JM, Hansen S. Concurrence of cystathioninuria, nephrogenic diabetes insipidus and severe anemia. *N Engl J Med* 1967;**276**:721–5.
28. Arthus MF, Lonergan M, Fujiwara TM, Bichet DG. Clinical and genetic approaches to the diagnosis of congenital polyuro-polydipsic syndromes, In: *The NDI Foundation 2004 Global Conference*Arizona: Phoenix; 2004. p. 55.
29. Spanakis E, Milord E, Gragnoli C. AVPR2 variants and mutations in nephrogenic diabetes insipidus: review and missense mutation significance. *J Cell Physiol* 2008;**217**:605–17.
30. Knops NB, Bos KK, Kerstjens M, van Dael K, Vos YJ. Nephrogenic diabetes insipidus in a patient with L1 syndrome: a new report of a contiguous gene deletion syndrome including L1CAM and AVPR2. *Am J Med Genet A* 2008;**146A**:1853–8.
31. Fujimoto M, Imai K, Hirata K, Kashiwagi R, Morinishi Y, Kitazawa K, et al. Immunological profile in a family with nephrogenic diabetes insipidus with a novel 11 kb deletion in AVPR2 and ARHGAP4 genes. *BMC Med Genet* 2008;**9**:42.
32. Bichet DG, Birnbaumer M, Lonergan M, Arthus M-F, Rosenthal W, Goodyer P, et al. Nature and recurrence of AVPR2 mutations in X-linked nephrogenic diabetes insipidus. *Am J Hum Genet* 1994;**55**:278–86.
33. Hobbs HH, Russell DW, Brown MS, Goldstein JL. The LDL receptor locus in familial hypercholesterolemia: mutational analysis of a membrane protein. *Annu Rev Genet* 1990;**24**:133–70.
34. Barak LS, Salahpour A, Zhang X, Masri B, Sotnikova TD, Ramsey AJ, et al. Pharmacological characterization of membrane-expressed human trace amine-associated receptor 1 (TAAR1) by a bioluminescence resonance energy transfer cAMP biosensor. *Mol Pharmacol* 2008;**74**:585–94.
35. Morello JP, Salahpour A, Laperrière A, Bernier V, Arthus M-F, Lonergan M, et al. Pharmacological chaperones rescue cell-surface expression and function of misfolded V2 vasopressin receptor mutants. *J Clin Invest* 2000;**105**:887–95.

36. Bernier V, Lagace M, Lonergan M, Arthus MF, Bichet DG, Bouvier M. Functional rescue of the constitutively internalized V2 vasopressin receptor mutant R137H by the pharmacological chaperone action of SR49059. *Mol Endocrinol* 2004;**18**:2074–84.
37. Hermosilla R, Oueslati M, Donalies U, Schonenberger E, Krause E, Oksche A, et al. Disease-causing V(2) vasopressin receptors are retained in different compartments of the early secretory pathway. *Traffic* 2004;**5**:993–1005.
38. Wuller S, Wiesner B, Loffler A, Furkert J, Krause G, Hermosilla R, et al. Pharmacochaperones post-translationally enhance cell surface expression by increasing conformational stability of wild-type and mutant vasopressin V2 receptors. *J Biol Chem* 2004;**279**:47254–63.
39. Robben JH, Deen PM. Pharmacological chaperones in nephrogenic diabetes insipidus: possibilities for clinical application. *BioDrugs* 2007;**21**:157–66.
40. Schoneberg T, Schulz A, Biebermann H, Hermsdorf T, Rompler H, Sangkuhl K. Mutant G-protein-coupled receptors as a cause of human diseases. *Pharmacol Ther* 2004;**104**:173–206.
41. Romisch K. A cure for traffic jams: small molecule chaperones in the endoplasmic reticulum. *Traffic* 2004;**5**:815–20.
42. Conn PM, Janovick JA. Trafficking and quality control of the gonadotropin releasing hormone receptor in health and disease. *Mol Cell Endocrinol* 2009;**299**:137–45.
43. Tamarappoo BK, Verkman AS. Defective aquaporin-2 trafficking in nephrogenic diabetes insipidus and correction by chemical chaperones. *J Clin Invest* 1998;**101**:2257–67.
44. Kunchaparty S, Palcso M, Berkman J, Velazquez H, Desir GV, Bernstein P, et al. Defective processing and expression of thiazide-sensitive Na-Cl cotransporter as a cause of Gitelman's syndrome. *Am J Physiol* 1999;**277**:F643–9.
45. Peters M, Ermert S, Jeck N, Derst C, Pechmann U, Weber S, et al. Classification and rescue of ROMK mutations underlying hyperprostaglandin E syndrome/antenatal Bartter syndrome. *Kidney Int* 2003;**64**:923–32.
46. Hayama A, Rai T, Sasaki S, Uchida S. Molecular mechanisms of Bartter syndrome caused by mutations in the BSND gene. *Histochem Cell Biol* 2003;**119**:485–93.
47. Chillaron J, Estevez R, Samarzija I, Waldegger S, Testar X, Lang F, et al. An intracellular trafficking defect in type I cystinuria rBAT mutants M467T and M467K. *J Biol Chem* 1997;**272**:9543–9.
48. Bonnardeaux A, Bichet DG. Inherited disorders of the renal tubule. In: Brenner BM, editor. *Brenner & rector's the kidney*. Philadelphia: Saunders; 2004. p. 1697–741.
49. Perucca J, Bichet DG, Bardoux P, Bouby N, Bankir L. Sodium excretion in response to vasopressin and selective vasopressin receptor antagonists. *J Am Soc Nephrol* 2008;**19**:1721–31.
50. Lomas DA, Evans DL, Finch JT, Carrell RW. The mechanism of Z alpha 1-antitrypsin accumulation in the liver. *Nature* 1992;**357**:605–7.
51. Lawless MW, Greene CM, Mulgrew A, Taggart CC, O'Neill SJ, McElvaney NG. Activation of endoplasmic reticulum-specific stress responses associated with the conformational disease Z alpha 1-antitrypsin deficiency. *J Immunol* 2004;**172**:5722–6.
52. Cohen FE, Kelly JW. Therapeutic approaches to protein-misfolding diseases. *Nature* 2003;**426**:905–9.
53. Ulloa-Aguirre A, Janovick JA, Brothers SP, Conn PM. Pharmacologic rescue of conformationally-defective proteins: implications for the treatment of human disease. *Traffic* 2004;**5**:821–37.
54. Barak LS, Oakley RH, Laporte SA, Caron MG. Constitutive arrestin-mediated desensitization of a human vasopressin receptor mutant associated with nephrogenic diabetes insipidus. *Proc Natl Acad Sci USA* 2001;**98**:93–8.
55. Bernier V, Morello JP, Zarruk A, Debrand N, Salahpour A, Lonergan M, et al. Pharmacologic chaperones as a potential treatment for X-linked nephrogenic diabetes insipidus. *J Am Soc Nephrol* 2006;**17**:232–43.
56. Glover DJ, Lipps HJ, Jans DA. Towards safe, non-viral therapeutic gene expression in humans. *Nat Rev Genet* 2005;**6**:299–310.

57. Sangkuhl K, Schulz A, Rompler H, Yun J, Wess J, Schoneberg T. Aminoglycoside-mediated rescue of a disease-causing nonsense mutation in the V2 vasopressin receptor gene *in vitro* and *in vivo*. Hum Mol Genet 2004;**13**:893–903.
58. Mankin AS, Liebman SW. Baby, don't stop!. Nat Genet 1999;**23**:8–10.
59. Feldman BJ, Rosenthal SM, Vargas GA, Fenwick RG, Huang EA, Matsuda-Abedini M, et al. Nephrogenic syndrome of inappropriate antidiuresis. N Engl J Med 2005;**352**:1884–90.
60. Decaux G, Vandergheynst F, Bouko Y, Parma J, Vassart G, Vilain C. Nephrogenic syndrome of inappropriate antidiuresis in adults: high phenotypic variability in men and women from a large pedigree. J Am Soc Nephrol 2007;**18**:606–12.
61. Soule S, Florkowski C, Potter H, Pattison D, Swan M, Hunt P, et al. Intermittent severe, symptomatic hyponatraemia due to the nephrogenic syndrome of inappropriate antidiuresis. Ann Clin Biochem 2008;**45**:520–3.
62. Marcialis MA, Faa V, Fanos V, Puddu M, Pintus MC, Cao A, et al. Neonatal onset of nephrogenic syndrome of inappropriate antidiuresis. Pediatr Nephrol 2008;**23**:2267–71.
63. Kalenga K, Persu A, Goffin E, Lavenne-Pardonge E, van Cangh PJ, Bichet DG, et al. Intra-familial phenotype variability in nephrogenic diabetes insipidus. Am J Kidney Dis 2002;**39**:737–43.
64. Kocan M, See HB, Sampaio NG, Eidne KA, Feldman BJ, Pfleger KD. Agonist-independent interactions between beta-arrestins and mutant vasopressin type II receptors associated with nephrogenic syndrome of inappropriate antidiuresis. Mol Endocrinol 2009;**23**:559–71.
65. Tesmer JJ, Sunahara RK, Gilman AG, Sprang SR. Crystal structure of the catalytic domains of adenylyl cyclase in a complex with Gsalpha.GTPgammaS. Science 1997;**278**:1907–16.
66. Mouillac B, Chini B, Balestre MN, Elands J, Trumpp-Kallmeyer S, Hoflack J, et al. The binding site of neuropeptide vasopressin V1a receptor. Evidence for a major localization within transmembrane regions. J Biol Chem 1995;**270**:25771–7.

Calcium-Sensing Receptor and Associated Diseases

GEOFFREY N. HENDY,[*]
VITO GUARNIERI,[†]
AND LUCIE CANAFF[*]

[*]*Departments of Medicine, Physiology, and Human Genetics, McGill University and Calcium Research Laboratory and Hormones and Cancer Research Unit, Royal Victoria Hospital, Montreal, Quebec, Canada H3A 1A1*

[†]*Medical Genetics Service, Hospital "Casa Sollievo della Sofferenza," Instituto di Recovero e Cura a Carattere Scientifico, San Giovanni Rotondo (Foggia), Italy*

I. Calcium Homeostasis	32
II. CASR and Diseases	33
III. CASR is a Family C GPCR	34
IV. Human CASR	34
V. Orthosteric Agonists	35
VI. Allosteric Modifiers	36
A. Positive Allosteric Modulators (Calcimimetics)	36
B. Negative Allosteric Modulators (Calcilytics)	36
VII. Structure and Function	37
A. Asparagine-Linked Glycosylation	37
B. Venus-Flytrap Domain	37
C. Dimerization	39
D. Ligand-Binding Sites in the VFT	39
E. Cysteine-Rich Domain	40
F. Peptide Linker	41
G. Extracellular Loops	42
H. Transmembrane Domain and Allosteric Modifiers	42
I. Intracellular Loops	44
J. COOH-Terminal Tail	44
VIII. Receptor Downregulation and Protein Kinase C	45
IX. Receptor-Activity-Modifying Proteins and CASR Trafficking	45
X. Ubiquitination and Conformational Checkpoint in CASR Processing	46
XI. CASR and Overview of Signaling Pathways	46
XII. CASR and the Parathyroid	48
XIII. CASR and the Renal Tubule	50
XIV. Disorders Associated with CASR (Table I)	50
A. Familial Hypocalciuric Hypercalcemia	50
B. Neonatal Severe Hyperparathyroidism	51
C. Molecular Genetics of FHH and NSHPT	52

	D. Autosomal Dominant Hypocalcemia.......	52
	E. Bartter's Syndrome Subtype V..........	53
XV.	CASR Mutation Repertoire.................	54
	A. CASR-Inactivating Mutations (Table II)	54
	B. CASR-Activating Mutations (Table III).	59
XVI.	Autoantibodies and the CASR..............	66
XVII.	CASR Polymorphisms.......................	66
XVIII.	Altered Expression of CASR and Disease..	71
	A. Hyperparathyroidism...................	71
	B. Hypercalciuria........................	72
	C. CASR and Cancer.......................	72
	D. Proinflammatory Cytokines.............	73
XIX.	CASR Allosteric Modifiers in the Clinic..	73
XX.	Summary..................................	73
	References...............................	74

The calcium-sensing receptor (CASR) is expressed in parathyroid hormone (PTH)-secreting cells of the parathyroid gland and cells lining the renal tubule. The activated CASR modulates intracellular signaling pathways altering PTH secretion and renal cation and water handling. Inherited abnormalities of the *CASR* gene give rise to a variety of disorders of mineral ion homeostasis. Heterozygous loss-of-function mutations cause familial (benign) hypocalciuric hypercalcemia (FHH) in which the lifelong mild hypercalcemia is generally asymptomatic. Homozygous inactivating mutations give rise to neonatal severe hyperparathyroidism (NSHPT) with extreme hypercalcemia and marked skeletal changes. Heterozygous activating mutations of the CASR cause autosomal dominant hypocalcemia (ADH) that may be asymptomatic or present with seizures in the neonatal period or childhood or later in life. Phenocopies of FHH or ADH are due to circulating CASR inactivating or activating autoantibodies, respectively. The CASR is the target of small molecule allosteric modifiers, either activators, calcimimetics, or inhibitors, calcilytics.

I. Calcium Homeostasis

Maintenance of blood levels of ionized calcium within a narrow normal range is of critical importance for control of neuronal excitability, muscle contraction, hormone secretion, and coagulation. The cell-surface calcium-sensing receptor (CASR) is the "calciostat" that orchestrates this systemic calcium homeostasis.[1] The CASR is expressed abundantly in the parathyroid hormone (PTH) producing chief cells of the parathyroid gland, the calcitonin-producing C-cells of the thyroid, and the cells lining the kidney tubule. The seven-transmembrane (7-TM) G-protein-coupled receptor (GPCR) senses small changes in circulating calcium concentration and modulates intracellular signaling pathways that alter PTH and calcitonin secretion or renal cation handling, thereby restoring blood mineral ion levels to normal.

The relationship between extracellular ionized calcium and PTH concentration is represented by an inverse sigmoidal curve. The activity and/or expression level of the CASR dictates the so-called calcium set-point, defined as the extracellular calcium concentration at which PTH secretion from the parathyroid gland (or calcium reabsorption across the kidney tubule) is half-maximal.[2] Increases in extracellular calcium directly stimulate calcitonin secretion[3–5] and the CASR on breast ductal cells regulates calcium transport into milk.[6]

Other tissues in which the CASR is expressed where it is likely to play a mineral ion homeostatic role include, skeletal tissues[7–9]—osteoclasts and their precursors,[10] osteoblasts and their precursors, osteocytes, chondrocytes[11–13]—and placenta.[14] Tissues in which the CASR is expressed at lower levels and is likely to play roles unrelated to mineral ion homeostasis include neurons and glia of the brain,[15] keratinocytes,[16] vascular smooth muscle cells,[17] hematopoietic stem cells (in blood and bone marrow),[18] stomach,[19] intestine,[20] colon,[21] liver,[22] pancreas,[23] and others.[24]

II. CASR and Diseases

Inactivating mutations of the CASR gene cause familial hypocalciuric hypercalcemia (FHH: MIM# 145980) in which the lifelong hypercalcemia is generally asymptomatic—the disease is also known as familial benign hypercalcemia; while homozygosity manifests as neonatal severe hyperparathyroidism (NSHPT: MIM# 239200).[25–27] Infants with NSHPT may develop severe, symptomatic hypercalcemia with skeletal changes of unremitting hyperparathyroidism and the disorder can be life-threatening without treatment. Heterozygous activating mutations in the CASR cause autosomal dominant hypocalcemia (ADH: MIM# 601198), autosomal dominant hypoparathyroidism (MIM# 241400), or hypocalcemic hypercalciuria (MIM# 146200).[28–30] Some individuals with ADH may have mild hypocalcemia and relatively few symptoms. However, in some cases seizures can occur, and these often happen during febrile episodes due to intercurrent infections.[31,32]

III. CASR is a Family C GPCR

The CASR is a member of family C of the superfamily of GPCRs that comprise three different subfamilies that have ≥20% amino acid identity over their transmembrane domain (TMD) comprising the 7-TM-spanning region.[33] The GPCRs are also referred to as 7-TM receptors. Group 1 comprises the metabotropic glutamate receptors, mGluR 1–8, that are widely expressed in the central nervous system that bind and are activated by the excitatory neurotransmitter, glutamate. Group II contains the CASR, the vomeronasal receptors (VRs), and taste and odorant receptors. The VRs are expressed exclusively in the rodent vomeronasal sensory organ (VNO), which responds to pheromones to direct instinctive behavior. Group III comprises the gamma-aminobutyric acid ($GABA_B$) receptors that bind and activate the neuro-inhibitor, $GABA_B$. The widely expressed orphan receptor, GPRC6A, also belongs in family C, and is most similar to the CASR. GPRC6A binds and is activated by some L-α-amino acids and other ligands including calcium.

IV. Human CASR

The human *CASR* gene that maps to 3q13.3–21[34] spans ~103 kb and has eight exons.[35] Six exons (exons 2–7) encode the CASR protein of 1078 amino acids. In exon 7, there is usage of two different polyadenylation signal sequences (AATAAA) yielding either a short (177-nucleotide) or long (1304-nucleotide) 3′-untranslated region (UTR).[36,37] Exon 2 encodes 242 nucleotides of the 5′-UTR, followed by the translation initiation codon. Exons 1A and 1B encode alternative 5′-UTRs that splice to the common portion encoded by exon 2.[36,38] The transcriptional start sites (+1) of promoters P1 and P2 have been precisely mapped.[39] Promoter P1 has a TATA box at nucleotide −26 and a CCAAT box at −110 relative to the start site. Promoter P2 has Sp1 sites at the transcriptional start site. Functional vitamin D response elements (VDREs),[39] NF-κB elements,[40] and Glial Cells Missing (GCM) elements[41] are present in both promoters. Functional Stat1/3 elements are in promoter P1 and Sp1/3 elements are in promoter P2.[42]

The NH_2-terminal signal peptide that directs the nascent CASR polypeptide into the endoplasmic reticulum (ER), and is then removed, has 19 amino acids.[43,44] The extracellular domain (ECD) of the mature protein has ~600 amino acids comprised of a bilobed Venus-flytrap-like domain (VFT) that is connected by a cysteine-rich region and a peptide linker to the seven-transmembrane domain (TMD) with an intracellular COOH-terminal tail of 216 amino acids.[45] The CASR functions as a dimer with both covalent and noncovalent interaction between the VFTs of each monomer.[46] Ca^{2+} binds in the

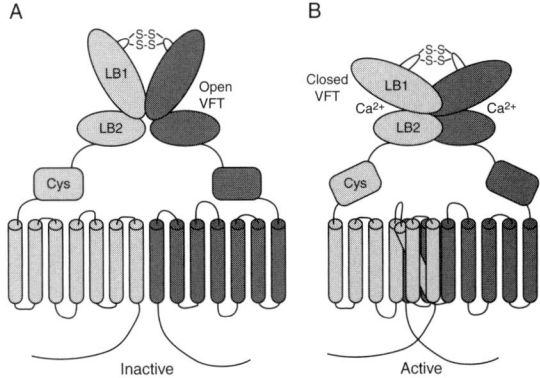

FIG. 1. Model of ligand activation of the CASR dimer at the cell surface. (A) Intermolecular disulfide bonds (in loop 2) link lobes 1 of each protomer (monomer) and in the absence of ligand maintain the VFT formed by lobes 1 and 2 in an open (inactive) conformation. (B) Ca^{2+} binds within the cleft between lobes 1 and 2 causing VFT closure and a rotation about the dimer interface with a change in configuration of loop 2 (the intermolecular disulfides). The conformation of the cysteine-rich region alters bringing about reconfiguration of some TM-helices such that intracellular loops contact G proteins triggering cell signaling.

cleft of each VFT causing the lobes to close on each other and the VFT to rotate and transfer a conformational change to the cysteine-rich region (Fig. 1). This results in movement of the membrane α-helices relative to one another thereby allowing G proteins to interact with intracellular loops and instigate cell signaling. The CASR couples largely via $G_{q/11}$ and G_i proteins although coupling via G_s has also been described[24,47,48] and activates or inhibits multiple intracellular signaling pathways.[49]

V. Orthosteric Agonists

The CASR is promiscuous and is activated by polycations, Ca^{2+}, Mg^{2+}, Ba^{2+}, Gd^{3+}, La^{3+}, Sr^{2+},[24] charged polyvalent molecules such as spermine and spermidine,[22,50] and β-amyloid peptides[51] and aminoglycoside antibiotics.[52] These agonists act by binding to sites in the ECD of the CASR.[53] Ionic strength,[54] pH and acid–base status,[55] and the L-α-amino acids, phenylalanine, tryptophan, and histidine[56–58] positively modulate sensitivity to Ca^{2+}. The CASR is likely to integrate diverse physiological signals, for example, those important for protein and mineral metabolism.[59]

VI. Allosteric Modifiers

A. Positive Allosteric Modulators (Calcimimetics)

In contrast to orthosteric agonists, small orally active compounds have been developed that bind within the TMD.[60,61] First-generation positive allosteric modulators are phenylalkylamine compounds derived from voltage-sensitive calcium channel blockers (Fig. 2). The compounds, NPS R-467 and R-568 act in a stereoselective way to enhance the sensitivity of the CASR to its orthosteric agonists while having no intrinsic activity of their own.[62,63] Second-generation calcimimetics such as cinacalcet have been developed with improved pharmacokinetic properties.[64] Calindol is similar in its properties to NPS R-568.[65,66] Calcimimetics act on parathyroid cells *in vitro* and *in vivo* to inhibit PTH secretion. Cinacalcet is in clinical use for treatment of uremic secondary hyperparathyroidism and parathyroid cancer.

B. Negative Allosteric Modulators (Calcilytics)

The first reported negative allosteric modulator was NPS 2143[67] and other calcilytics of different structure to NPS 2143 are Calhex 231[68] and Compound 1[69] (Fig. 2). Calcilytics antagonize the parathyroid cell CASR *in vitro* and *in vivo* and stimulate PTH secretion.[70]

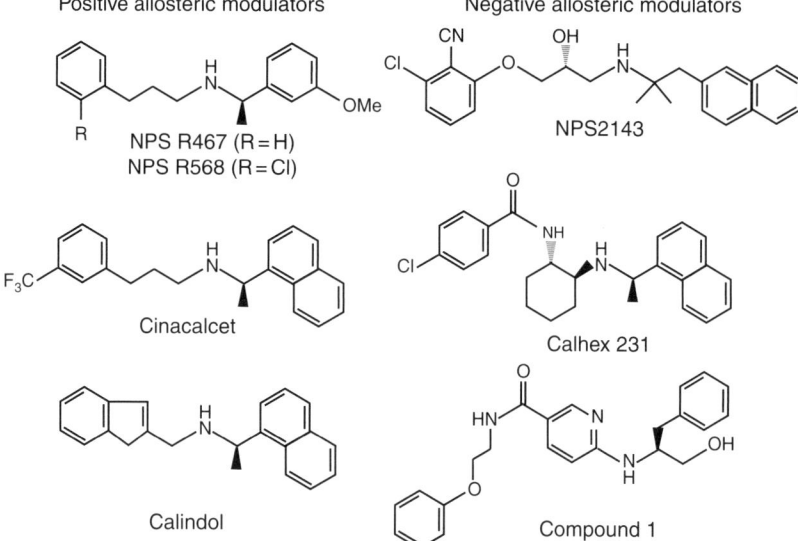

FIG. 2. Allosteric modulators of the CASR. Positive modulators: NPS R-467 and NPS R-568, Cinacalcet and Calindol. Negative modulators: NPS 2143, Calhex 231, and Compound 1.

VII. Structure and Function

Functional characterization of the CASR with respect to important structural features has been achieved most often by use of so-called heterologous systems in which a mammalian expression vector with a cDNA insert (often with an epitope tag) representing either wild-type or mutant CASR is transfected into human embryonic kidney (HEK) 293, or in a few cases, COS, cells.[71] In this review, HEK293 cells transfected with the CASR are designated as CASR-HEK cells. A variety of analyses are used to assess overall expression, expression at the plasma membrane, glycosylation and dimerization status, and signaling capability via several pathways.[46,72–75] Relative binding affinities of ligands are inferred indirectly by the responsiveness of cell signaling pathways. There is no Ca^{2+}-binding assay.

A. Asparagine-Linked Glycosylation

Western blot analysis of extracts of HEK293 cells transfected with CASR run under standard SDS–PAGE conditions reveals a variety of molecular species and the nature of these has been demonstrated by use of tunicamycin and endoglycosidase H (endo H) and peptide-N-glycosidase F (PNGase F) to interrogate the overall and the immature and mature N-linked glycosylation status, respectively, as well as more robust denaturation conditions to evaluate dimerization.[71] The species are 120 kDa (nonglycosylated monomer), 140 kDa (immature glycosylated monomer), 160 kDa (mature glycosylated monomer), and other species >280 kDa (dimers). N-linked glycosylation is required for expression of the CASR at the cell surface.[76] Of the 11 potential N-linked glycosylation sites on the ECD of the human CASR, eight sites (N-90, N-130, N-261, N-287, N-446, N-468, N-488, and N-541) are used, whereas the three remaining sites (N-386, N-400, and N-594) are not (Fig. 3). Glycosylation of at least three sites is critical for cell-surface expression, but glycosylation is not critical for signal transduction.[77]

B. Venus-Flytrap Domain

The ECDs of the CASR and the other Group C GPCRs have limited amino acid homology to bacterial periplasmic amino acid-binding proteins for which the crystallographic structures are known.[78] Like these binding proteins, the ECDs have a so-called VFT or clam shell structure consisting of two lobes each with α-helices and β-sheet folds connected by a hinge region of three interwoven strands. This has been confirmed by 3D analysis of crystals of the mGluR1 ECD in open (ligand-free) and closed (ligand-bound) conformations[79,80] and similar structures deduced by homology modeling for the CASR ECD (amino

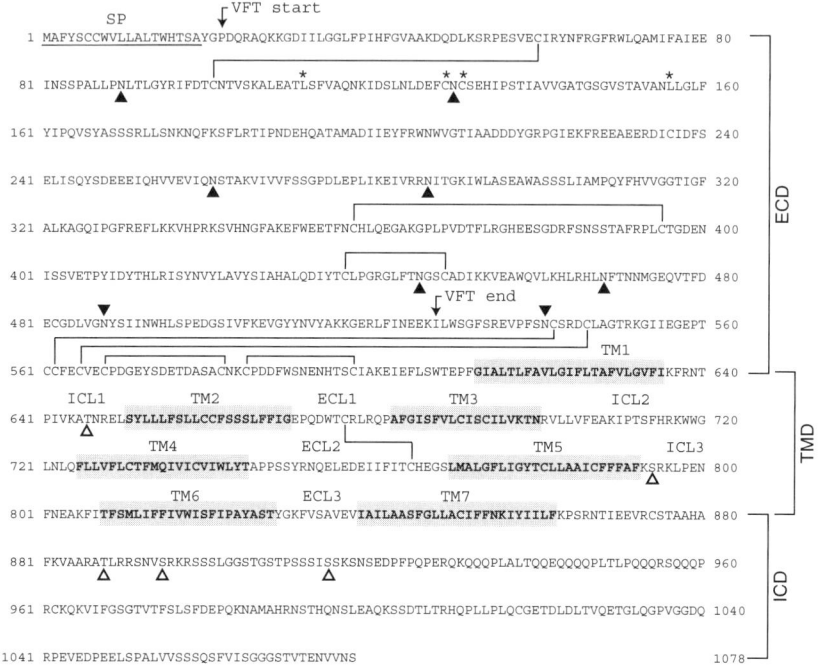

FIG. 3. Amino acid sequence (1–1078) and key features of the human CASR: SP, signal peptide (1–19) that is underlined; ECD, extracellular domain (20–612); TMD, transmembrane domain (613–862); ICD, intracellular domain (863–1078). In the ECD: VFT, Venus-flytrap domain, arrows indicate start (22) and end (528); intramolecular disulfide bonds, cysteine (C) residues joined by lines; functional N-linked glycosylation sites, filled arrowheads; asterisks (∗), leucine (L) residues and cysteine (C) residues contributing to hydrophobic and covalent intermolecular linkages, respectively, for dimer formation. In the TMD: transmembrane α-helices (TM1–TM7), bolded residues on shaded boxes; intracellular loops (ICL1–ICL3); extracellular loops (ECL1–ECL3); disulfide bond between ECL2 and ECL3, cysteine (C) residues joined by a line. PKC sites, open arrowheads (S or T) in TMD and ICD.

acids 23–528).[45,46,81] Some parts of the CASR ECD do not align with the bacterial protein structure and are represented by four protruding loops: loop 1 (50–59), loop 2 (117–136), loop 3 (365–385), and loop 4 (438–445). By study of engineered mutants, deletion of loop 1 leads to reduced activity of the receptor.[82] The VFT model predicts a rotation of one lobe relative to the other upon ligand binding which directly or indirectly modifies the conformation of the TMD leading to receptor activation. A cysteine-rich region from amino acids 542 to 598 may play a role in signal transduction, after ligand binding, from the more NH_2-terminal part of the ECD to the TMD.[83,84]

C. Dimerization

The CASR exists in a functional form as dimers on the cell surface of transfected cells and this is promoted in part through intermolecular disulfide bonds involving C129 and C131 in the ECD.[85,86] These cysteine residues are proposed to be located in loop 2 (117–136) that protrudes from one lobe of the VFT of each monomer (or protomer).[87] However, these covalent linkages are not needed for functional interactions between each CASR protomer within the dimer. In fact, removal of loop 2 leads to increased sensitivity to Ca^{2+}.[82] The CASR has at least two types of motifs mediating dimerization (and functional interactions); covalent interactions (intermolecular disulfide bonds) and noncovalent (hydrophobic) interactions.[88] Two leucine residues, L112 and L156, in the ECD are important for noncovalent dimerization. A receptor in which the leucines are mutated (L112S and L156S) still exists as a covalently linked dimer with higher affinity for calcium than the wild-type receptor. A combination of the four mutations, L112S, L156S, C129S, and C131S, reduces receptor dimerization and markedly inactivates the CASR. Dimerization through the ECD is essential for formation of a functional tertiary structure of the CASR.[89] With respect to the other cysteines in the VFT, it can be predicted, by homology modeling of the CASR structure on the mGluR ECD, that intramolecular disulfide bonds form between C60–C101, C358–C395, and C437–C449 (Fig. 3). Site-directed mutagenesis studies showed that each of these cysteines are critical for CASR expression and function and are likely to be critical for the VFT structure.[90] Intracellular CASR dimers were identified in rat kidney medulla extracts[86] and in CASR-HEK cells.[91] Constitutive dimerization was demonstrated within the ER by photobleaching fluorescence energy transfer microscopy.[92]

D. Ligand-Binding Sites in the VFT

Ca^{2+} activates the CASR at millimolar concentrations consistent with low-affinity binding and there is positive cooperativity suggesting multiple Ca^{2+} binding sites. Several of the 13 amino acids in the VFT of the mGluR1 involved in glutamate binding are identical or conservatively substituted in the CASR.[58] In one study, it was proposed that the Ca^{2+}-binding site in the CASR comprises polar residues—S170, D190, Q193, S296, and E297—directly involved in Ca^{2+} coordination, and an additional set of residues that contributes to the "coordination sphere" of the cation (F270, Y218, S147).[93] Another study identified three potential Ca^{2+}-binding sites in a modeled CASR structure.[94] To probe the intrinsic Ca^{2+}-binding properties of predicted sequences, two predicted continuous Ca^{2+}-binding sequences were individually engineered into a scaffold protein, the non-Ca^{2+}-binding protein, CD2. One of these sites is the same or similar to that described previously.[93] In further studies, three globular subdomains in the intact CASR structure have been identified each of which is predicted to contain two to three Ca^{2+}-binding sites[95] (Fig. 4).

Studies of the related mGluRs provide evidence that the VFT domains may function somewhat differently in different mGluRs. For heterodimers of mGluR5-$GABA_{B1}$ while closure of one mGluR VFT leads to partial activation, closure of both VFTs in the dimer is needed for full activity.[96] mGluR3 ECD crystal structures with ligands show that rather than changes in protein conformation the different agonists rearrange solvent molecules.[97] This contrasts with the VFT being closed versus open under glutamate-bound versus not bound conditions for mGluR1[79,80]—the model presently proposed for the CASR. There is a rotation of one monomer relative to the other about an axis perpendicular to the dimer interface bringing the COOH-termini of the two VFT monomers closer. Consistent with this model being relevant for the CASR, two different monoclonal antibodies against lobe 2 of the VFT either increase or decrease sensitivity to extracellular Ca^{2+} presumably by enhancing or inhibiting agonist-promoted VFT closure or rotation.[98]

E. Cysteine-Rich Domain

Between the VFT and the start of the TMD, the CASR has a Cys-rich region with nine conserved cysteines and if any one of these is mutated, expression and function of the CASR is impaired.[90] Deletion of the entire Cys-rich region preserves some cell-surface expression but Ca^{2+} activation is lost.[84] By homology modeling based upon the crystal structure of the mGluR3 ECD,[97] the Cys-rich region of the CASR has three β-sheets, each comprised of two antiparallel β-strands. Four pairs of cysteines, C542–C562, C546–C565, C568–C582, and C585–C598, form disulfide linkages. The remaining cysteine, C561, has been reported to be unpaired[90,99] although extrapolation from data of the mGluR3 structure suggests potential linkage between C236 (in lobe 2 of the VFT) and C561. Perhaps, in the CASR, this disulfide bond is particularly labile relative to the others. The exact role that the Cys-rich region plays in transmitting agonist-induced conformational changes in the VFT to the TMD is not known. Also, not all family C GPCRs have a Cys-rich domain; $GABA_B$ receptors do not.

The short region (529–541) between the end of the VFT and the Cys-rich region can be variable in length as a fully functioning splice variant having 10 additional amino acids has been identified[36] and engineered alterations in that region do not affect function.[99]

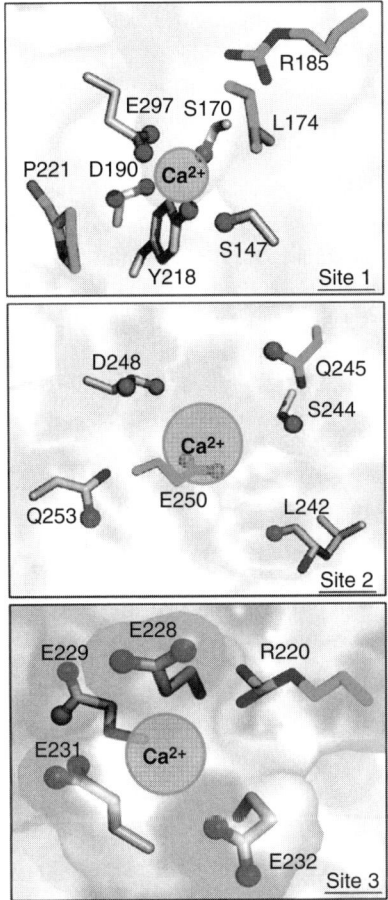

FIG. 4. Location of three predicted Ca^{2+}-binding sites in a subdomain (amino acids 132–300) of the ECD of the CASR. Site 1 is in the hinge region of the VFT structure formed by lobes 1 and 2 (see Fig. 1). Sites 2 and 3 are in the first half of lobe 2 (amino acids 215–253). Sites 4 and 5 (not shown) are clustered in the second half of lobe 1 (amino acids 350–400). (Redrawn with permission from Ref. 95.) (See Color Insert.)

F. Peptide Linker

A 14 amino acid linker (599–613) connects the Cys-rich region to the TMD. A similar linker is found in all family C GPCRs with the exception of the $GABA_B$ receptor. Small alterations in the length and composition of the linker impaired cell-surface expression and abrogated signaling. Alanine substitution of four conserved residues identified L606 as being critical for cell-surface expression and signaling. However, substitution of L606 with either Ile or Val led to activation. Therefore, it was suggested the linker region, in particular L606, plays a critical role in transmission of the active signal from ECD to TMD.[100]

G. Extracellular Loops

Mutation of any of three acidic residues in extracellular loop 2 (ECL2), D758, E759, and E767, increased sensitivity to Ca^{2+} activation of both full-length receptor and one lacking the ECD. Mutation of E837 in ECL3 reduced sensitivity to the calcimimetic NPS R-568. Mutation of other acidic residues had no effect. It was suggested that the three acidic residues in ECL2 maintain an inactive conformation of the TMD.[101]

The roles of charged amino acids and cysteines in the three extracellular loops were examined by alanine-scanning mutagenesis.[102] Mutation of C677 and C765 in ECL1 and ECL2, respectively, led to altered glycosylation, reduced cell-surface expression to 5% of wild type, and completely ablated phosphoinositol hydrolysis (PI). In these studies, replacement of charged amino acids produced only minor changes in receptor activation except for E767A and K831A in ECL2 and ECL3, respectively, that showed gain of function by increased apparent Ca^{2+} affinity. A double mutant E767K/K831E in which the positions of charged residues were exchanged had impaired cell-surface expression and no PI response to Ca^{2+}. Hence, the two cysteines form critical disulfide links and the side chains of E767 and K831 are likely to be involved with ionic interactions with other amino acids. It was suggested that the interactions are important for receptor folding and maintaining the TMD in an inactive conformation.

H. Transmembrane Domain and Allosteric Modifiers

Agonists and antagonists of rhodopsin-like family A GPCRs bind within the TMD. Similarly, the allosteric modulators of the CASR (a family C GPCR) bind within the TMD. Models of the CASR TMD have been constructed based on the crystal structure of bovine rhodopsin in the inactive form with retinal bound.[103] It should be noted that the TMDs of the family A GPCRs like rhodopsin and family C GPCRs like CASR have no significant sequence identity. The sizes and sequences of the extracellular loops and intracellular loops of family A and C receptors are different with the exception of the two cysteines that form a disulfide bond between ECL1 and ECL2. However, the models have apparently been useful in identifying contact sites between residues in the CASR TMD and allosteric modulators.[104–108] It is important to note that assignments of the residues that comprise the TM-helices changed somewhat after the introduction of the rhodopsin model. This had the effect of reducing the sizes of some of the intracellular and extracellular loops. This was most apparent for ICL1, ECL1, ICL3, and ECL3. An instructive comparison can be made by reference to Fig. 2^{109} for the original TM assignments and to Fig. 1^{45} for the rhodopsin-type assignments. The present review is using the original TM assignments.

Two calcimimetic allosteric modulators, NPS R-568 and Calindol, that bind the TMD of the CASR have been reported to potentiate Ca^{2+} activation without independently activating the wild-type receptor. The ability of Calindol to activate a CASR construct (T903-Rhoc) in which the ECD and carboxyl-terminal tail have been deleted to produce a rhodopsin-like TMD was tested. Although Calindol has little or no agonist activity in the absence of extracellular Ca^{2+} for the ECD-containing wild-type or carboxyl-terminal deleted receptors, it acts as a strong agonist of the T903-Rhoc. Ca^{2+} alone displays little or no agonist activity for the CASR TMD, but potentiates the activation by the calcimimetic.[104]

The docking of Calhex 231, a negative allosteric modulator, has been evaluated with the 3D model of the CASR TMD. In the model, Glu-837 (TM7) anchors the two nitrogen atoms of Calhex 231 and locates the aromatic moieties in two adjacent hydrophobic pockets delineated by TMs 3, 5, and 6 and TMs 1, 2, 3, and 7, respectively. Two receptor mutations, F684A (TM3) and E837A (TM7) caused a loss of the ability of Calhex231 to inhibit Ca^{2+}-induced cell signaling. Three other mutations, F688A (TM3), W818A (TM6), and I841A (TM7), produced a marked increase in the IC_{50} of Calhex 231 for the Ca^{2+}-response whereas L776A (TM5) and F821A (TM6) led to a decrease in the IC_{50}. The data validate the proposed model for interaction of Calhex 231 with the TMDs of the CASR. Residues at the same positions have been shown to delimit the antagonist-binding cavity of diverse GPCRs [105]

The same group modeled the binding pocket of the two calcimimetics, NPS R-568 and Calindol, and compared the findings with those for calcilytics. There are (subtle) differences between the binding of calcimimetics and calcilytics. Some mutations have no effect on calcimimetics but affect the binding of calcilytics in TM3 and TM5 suggesting that the binding pockets of the positive and allosteric modulators are partially overlapping but not identical.[106] Residues W818, F821, E837, and I841 that are located in transmembrane helices, TM6 and TM7, were involved in the binding pocket for both calcimimetics and calcilytics. Additional residues located in TM3 (R680, F684, and F688) were involved in the recognition of calcilytics.[106]

A different negative allosteric modulator, compound 1, retains activity against the E837A mutant that lacks a response to other positive and negative modulators.[108] The related compound, JKJ05, acts as a negative modulator on the wild-type receptor but as a positive modulator on the E837A mutant. This activity is dependent upon the primary amine in JKJ05 that interacts with the acidic E767 in the 7-TM model generated in this study.[108]

The TMD has six prolines that cause kinks in the transmembrane helices that are likely to be functionally important. An engineered P823A mutation in TM6 was well expressed but demonstrated markedly reduced activation by Ca^{2+}.[110]

I. Intracellular Loops

Mutagenesis of residues within either ICL2 or ICL3 led to reduced cell signaling monitored by inositol trisphosphate (IP_3) production in response to elevations in extracellular Ca^{2+}. A signaling deficient phenotype was observed for F707A in ICL1, and L798A, F802A, and E804A in ICL2.[111] These studies were conducted with bovine CASR (that has 1085 residues compared to the 1078 of the human) and the equivalent residues in the human CASR would be F706, L797, F801, and E803.

J. COOH-Terminal Tail

The COOH-terminal intracellular tail contributes to several properties of the receptor; intracellular signaling, level of cell-surface expression, and rate of desensitization.

Deletion of the majority of the carboxyl terminus (889–1078) is compatible with normal processing, cell-surface expression, and signal transduction.[112] However, mutants truncated at 706 and 802 within the second and third intracellular loops, respectively, lack a signaling response to extracellular Ca^{2+} and are not properly glycosylated and fail to reach the cell surface. Although these mutants did not exhibit a mature glycosylated monomeric species, it can be noted that dimers were formed. Mutants truncated at 888 and 903 within the carboxyl terminus were equivalent to the wild type in all assays. A small region between residues 874 and 888 is critical for normal signal transduction. Mutants truncated at 865 and 874 had no response to extracellular Ca^{2+} despite only a slight (~25%) reduction in cell-surface expression. Full-length CASR mutants with residues between 874 and 888 substituted by alanines showed either no (875A, 876A, 879A) or reduced (881A–883A) calcium response at levels of cell-surface expression equivalent to wild type.

In a separate study, there were no phospholipase C (PLC) responses to high extracellular Ca^{2+} in cells expressing CASR 1–866 (with a COOH-tail of three residues) although receptors were expressed at the plasma membrane.[113] The residues between S866 and V895 were scanned with tandem-Ala and single site mutagenesis. Two point mutants H880A and F882A showed 50–70% reductions in high extracellular Ca^{2+}-induced IP_3 production. Levels of expression and glycosylation of the mutants were comparable with wild-type CASR but the mutant receptors were retained in intracellular organelles and colocalized with an ER marker. It was suggested that the signaling defects were likely due to defective trafficking to the cell surface. Modeling indicated a putative α-helical structure (15 amino acids) between 877 and 891. The data suggested that specific amino acids, and possibly unique secondary structure, are required for efficient targeting of the CASR to the cell surface. Note that as these studies were done with bovine CASR cDNA, the equivalent residue numbers for human CASR are one less than those given earlier.

VIII. Receptor Downregulation and Protein Kinase C

The CASR is desensitized to a limited extent by repeat exposure to agonists and phosphorylation either by GPCR kinases and recruitment of β-arrestin[114] or by protein kinase C (PKC) and reduced capacity to activate PLC.[115] The activated CASR stimulates PLC that mobilizes intracellular Ca^{2+} stores, on the one hand, and activates PKC, on the other. The CASR has five potential PKC sites in its intracellular domains (IC loops and intracellular tail) and phosphorylation of T888 (the most important site), S895, and S915 in the intracellular tail maximally inhibits Ca^{2+} mobilization.[116] Truncating the receptor at T888 has a negative effect on release of intracellular Ca^{2+} stores without affecting the activation of Ca^{2+} influx. Although not shown directly, the PKC phosphorylation of the CASR was suggested to prevent interaction with G protein subtypes critical for releasing Ca^{2+} stores and this was independent of activation of Ca^{2+} influx.[117]

IX. Receptor-Activity-Modifying Proteins and CASR Trafficking

Receptor-activity-modifying proteins (RAMPs) associate with some of the family B GPCRs, and this can be required for their functional expression at the cell surface. The particular RAMP (1, 2 or 3) can dictate the ligand specificity of particular receptors. A transfected pH-sensitive GFP superecliptic pHluorin (SEP)-CASR was retained in the ER in COS7 cells that do not contain endogenous RAMPs. SEP-CASRs were delivered to the plasma membrane in HEK293 cells that do express RAMP1.[118] Coexpression of RAMP1 or RAMP3, but not RAMP2, in COS7 cells was sufficient to target the CASR to the cell surface. These and further experiments supported the notion that association of RAMPs is necessary and sufficient to transfer the immature CASR from the ER to the Golgi where it becomes fully glycosylated. The basis for these studies is that while HEK293 cells widely used for CASR transfection studies do express the required RAMPs, COS7 cells do not. However, it can be noted that other investigators have successfully used COS7 cells for CASR transfection studies.[119,120] Further studies are required to assess the physiological significance of these findings.

X. Ubiquitination and Conformational Checkpoint in CASR Processing

CASR is ubiquitinated by the E3 ligase dorfin and degraded via an ER-associated degradation pathway[121] and evidence was provided for conformational or functional checkpoint in CASR biogenesis.[122] The finding of stabilization of a subset of the loss-of-function CASR mutants tested but not gain-of-function CASR mutants by the proteasome inhibitor MG132 suggested that receptor sensitivity to calcium influences susceptibility to proteasomal degradation. The allosteric activator NPS R-568 and antagonist NPS 2143 were used to promote the active and inactive conformations of the CASR, respectively. Overnight treatment with NPS R-568 or NPS 2143 differentially regulated maturation and cell-surface expression of wild-type CASR, directly affecting maximal signaling responses. NPS R-568 rescued expression of loss-of-function CASR mutants, increasing plasma membrane expression and ERK1/2 phosphorylation in response to 5 mM Ca^{2+}.[122] Further studies have confirmed that a subset (but not all) of additional missense CASR inactivating mutants, including some that are already expressed at the cell surface, show enhanced responsiveness to extracellular Ca^{2+} after treatment with the "pharmacochaperone," NPS R-568.[123,124] Disordered calcium homeostasis caused by some CASR mutations may result from altered receptor biogenesis independent of receptor function (a protein folding disorder). Allosteric modulators may not only alter CASR sensitivity to calcium (and signaling) but also modulate receptor expression.

XI. CASR and Overview of Signaling Pathways

The CASR is a pleiotropic GPCR that can couple to more than one type of G protein. The most well-characterized interactions are with $G_{q/11}$ and G_i, but coupling to $G_{12/13}$, G_o, and G_s occurs in some cell types. Stimulated CASR couples to G_q (pertussis toxin-insensitive) causing PLC-mediated 1,2-diacylglycerol (DAG) formation, and IP_3 formation with intracellular Ca^{2+} mobilization and couples to G_i (pertussis toxin-sensitive) causing inhibition of cAMP formation (Fig. 5). By increasing the intracellular Ca^{2+} concentration and phospholipid metabolites such as DAG, the activated CASR activates the serine/threonine kinase PKC. Both conventional and atypical isoforms of PKC can be involved. Via PLC and PKC activation, the activated CASR also stimulates phospholipase A_2 (PLA_2) with arachidonic acid production and phospholipase D (PLD) with phosphatidic acid formation. The mitogen-activated protein kinase (MAPK) family includes the proline-directed serine/threonine

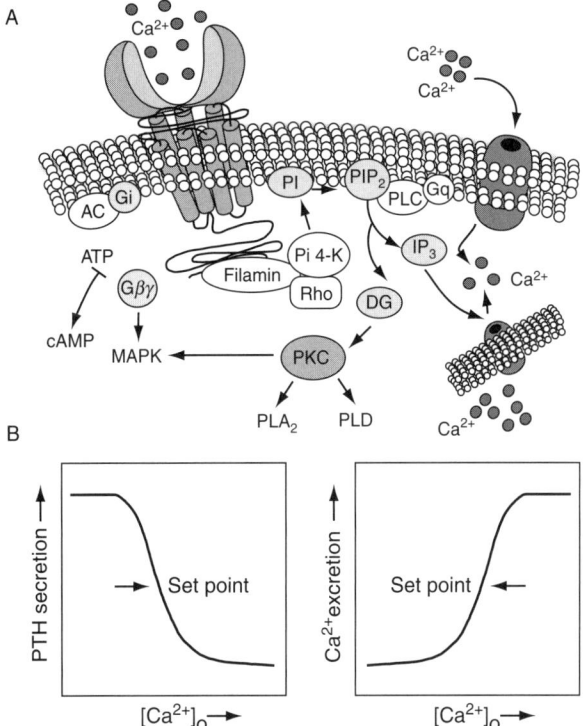

FIG. 5. (A) Signaling pathways activated by the CASR. (B) The CASR controls the relationship between the extracellular calcium concentration $[Ca^{2+}]_o$ and PTH secretion on the one hand and urinary calcium excretion on the other. The set-point is the $[Ca^{2+}]_o$ at which PTH secretion or calcium excretion is half-maximal. (See Color Insert.)

protein kinases ERK1 (p44) and ERK2 (p42), p38 and stress-activated c-Jun N-kinases (JNK). Both $G_{i/o}$ and $G_{q/11}$ coupling have been implicated in ERK1 and ERK2 phosphorylation by the activated CASR. The stimulated CASR can also activate p38 and JNK.[49]

When the cytoplasmic calcium responses of individual cells are examined by single cell fluorescence recording, small increments in extracellular calcium sustain intracellular Ca^{2+} oscillations that decay to a nonoscillatory plateau.[125] Besides the so-called type 1 agonists (e.g., calcium itself), the allosteric type 2 agonists including the calcimimetic NPS R-568 and aromatic amino acids, L-phenylalanine and L-tryptophan, trigger CASR-activated intracellular Ca^{2+} oscillations although there are some quantitative and qualitative differences between the two type of agonists. Aromatic amino acids produce transient oscillations whereas extracellular Ca^{2+} produces sinusoidal oscillations perhaps because calcium is both agonist and permeates the cell to influence the oscillations. In some cells, the extracellular calcium activation of oscillations in intracellular Ca^{2+}.[126–128] occurs through PLC activation and the Ca^{2+} permeable channel, TRPC1. The signaling determinants involved in CASR-generated intracellular Ca^{2+} oscillations appear to lie within amino acids 868–888 of the intracellular tail.[126,127]

XII. CASR and the Parathyroid

The precise mechanisms whereby activation of the CASR inhibits PTH secretion and synthesis and parathyroid cell proliferation are not known. Activation of the $G_{q/11}$ pathway appears critical as mice with parathyroid knockout of both G_q and G_{11} have a phenotype of severe hyperparathyroidism mimicking that of mice (and humans) with complete deletion of the *Casr* (*CASR*).[129] Confirmation that the G_q-coupled pathway is important in part comes from the finding that transgenic mice specifically overexpressing a dominant-negative $G\alpha_q$ loop minigene in parathyroid chief cells had increased PTH mRNA and serum PTH levels and parathyroid gland size. The abnormalities that reflected reduced sensitivity of the parathyroid gland to calcium were similar to those of heterozygous *Casr* knockout mice.[130]

In parathyroid cells, agonist activation of CASR stimulates activation of phospholipases, PLC, PLA_2, and PLD.[131] CASR activation of PLC involves a pertussis toxin-insensitive pathway via the G_q and G_{11} proteins and also a pertussis toxin-sensitive pathway via G_i isoforms. Activation of PLC leads to PLA_2 and PLD activation mediated by DAG stimulated PKC. At the parathyroid cell membrane, the CASR is bound directly or indirectly to calveolin-1 and resides within caveolae, flask-like invaginations of the plasma membrane.[132] Filamins are a family of nonmuscle actin binding proteins that cross-link actin into a three-dimensional structure. Filamins interact with a very large number of cellular proteins of diverse functions and act as signaling scaffolding molecules. Filamin-A interacts with caveolin-1 and also binds the COOH-terminal tail of the CASR.[133–136] This appears to facilitate the ability of the CASR to activate MAPKs such as ERK1/2. In heterologous systems, the interaction between CASR and filamin-A stabilizes the CASR and protects it from desensitization.[137]

Increases in extracellular Ca^{2+} cause transient increases in intracellular Ca^{2+} in parathyroid cells by activation of PLC[131] resulting in IP_3-mediated release of intracellular Ca^{2+} stores.[73] Sustained increases in intracellular Ca^{2+} occur through an influx pathway that may involve a nonselective cation channel.[138,139] High extracellular Ca^{2+} inhibits agonist-stimulated cAMP in parathyroid cells, an action thought to involve inhibition of adenylate cyclase by an isoform of G_i.[140] The CASR stimulates the activity of MAPK (ERK1/2 and p38) through both a PKC-dependent pathway downstream of G_q-mediated activation of PLC then PKC, and via tyrosine kinases that utilize $\beta\gamma$ subunits released upon activation of G_i.[141,142] Activated MAPK stimulates PLA_2, releasing free arachidonic acid that is then further metabolized.[141] Previous studies showed that the lipoxygenase pathway of arachidonate metabolism is critical for inhibition of PTH secretion.[143] Addition of 12- and 15-hydroxyeicosatetranoic acid (HETE) and the hydroperoxyeicosatetranoic acid (PHETE) inhibits PTH secretion from parathyroid cells.[144] While there is evidence for a pathway starting with activated CASR, MEK/MAPK, and arachidonate metabolites, the further links to PTH release are not known.

There is an accumulation of F-actin, caveolin-1, filamin-A, PTH-containing vesicles, and the CASR at the apical secretory pole of parathyroid cells under high extracellular Ca^{2+} concentrations.[145] Actin polymerization is critical for inhibition of PTH release as incubation with actin depolymerizing agents promoted PTH secretion at both low and high extracellular Ca^{2+} concentrations. It is suggested that activated CASR may influence the phosphorylation status of filamin-A, thereby controlling actin polymerization and PTH release. Further studies are required to identify the critical intermediary signaling molecules activated by the CASR in this mechanism.

Calpains are a family of proteases that recognize bonds between domains and cleave at nonspecific amino acid sequences. Calpains hydrolyze target proteins in a restricted manner to produce large fragments retaining intact domains of the original protein. In caveolae preparations from bovine parathyroid cells incubated with low extracellular Ca^{2+} (0.5 mM), inactive heterodimeric forms of m-calpain are present and raising the extracellular Ca^{2+} to 3 mM decreases the amount of m-calpain with an increase in CASR protein and phosphorylated PKC-α and -β.[146] Localization of m-calpain within caveolae may maintain the enzyme in an inactive state whereas in the long term the CASR may be degraded by m-calpain.

Elevations in extracellular Ca^{2+} inhibit PTH mRNA levels[147] and calcimimetics that sensitize the parathyroid cell CASR to calcium have the same effect.[63] The regulation of PTH synthesis by the activated CASR occurs by a posttranscriptional mechanism involving destabilization of PTH mRNA.[148] The importance of the CASR in the tonic control of parathyroid cell proliferation is exemplified by the loss of this control in humans or mice lacking the *CASR* gene in which marked parathyroid hyperplasia ensues.[149] Treatment of rats with renal impairment with calcimimetics prevents the secondary parathyroid hyperplasia that would normally occur.[150,151] Supplementation of mouse models lacking components of the vitamin D endocrine system (either ligand or receptor) with calcium reduces the parathyroid hyperplasia.[152] The pathways involved whereby CASR controls parathyroid proliferation are not known.

XIII. CASR and the Renal Tubule

CASR is expressed along the length of the kidney tubule. The CASR is present on the basolateral side of the cortical thick ascending limb (CTAL), at high levels, and the distal convoluted tubule (DCT), that play critical roles in hormone-regulated calcium and magnesium reabsorption.[153] The CASR is also expressed at the base of the microvilli of the proximal tubular brush border, the basolateral side of the medullary thick ascending limb (MTAL) tubular cells, and the luminal side of the epithelial cells of the inner medullary collecting duct (IMCD).[154] In the proximal tubule, the activated CASR inhibits PTH-stimulated phosphaturia and also is likely to mediate the direct calcium control of the proximal tubular 25-hydroxyvitamin D-1α-hydroxylase.[39,155]

In the CTAL, high levels of peritubular but not tubular calcium inhibit calcium and magnesium reabsorption in isolated tubule segments,[156] an effect likely to be mediated by the basolateral CASR on paracellular ion transport. Mechanistically, this occurs via CASR inhibition of the hormone-stimulated cAMP levels thereby reducing the activity of the Na, K, 2Cl cotransporter on the one hand and stimulating arachidonic acid that is metabolized by a P450 pathway on the other[157] to inhibit other channels and cotransporters critical for calcium and magnesium reabsorption (see later section on Bartter's syndrome type V). Extracellular calcium modulates the expression of key genes in the DCT responsible for transcellular calcium uptake such as those for the apical uptake channel, TRPV5, calbindin D, the basolateral calcium pump, PMCA2b, and exchanger, NCX1.[158–160] While these changes occur in part via alterations in circulating PTH and 1,25(OH)$_2$D, direct effects of calcium also play a role.[161] The DCT is difficult to investigate *in situ* but studies with a mouse distal convoluted tubule (MDCT) cell line demonstrated expression of the CASR and that its activation with either calcium or magnesium inhibited hormone (PTH, calcitonin, glucagon, or arginine vasopressin)-stimulated cAMP levels.[162] In addition, the magnesium/calcium sensing inhibited the hormone-stimulated cation (magnesium) uptake in these cells.[163]

XIV. Disorders Associated with CASR (Table I)

A. Familial Hypocalciuric Hypercalcemia

Reports of a syndrome called familial benign hypercalcemia appeared some 40 years ago.[164,165] The clinical features were documented in several large kindreds in which not only the generally asymptomatic nature of the lifelong hypercalcemia[166] but also the abnormal renal handling of calcium[167] was noted. In this familial (benign) hypocalciuric hypercalcemia (FHH) syndrome the hypercalcemia is usually mild, with serum calcium levels no more than 10% above the upper limit of normal, although a few families exhibit higher serum calcium concentrations.[168] The symptoms and complications seen in patients with other forms of hypercalcemia are not a part of FHH. Circulating vitamin D metabolite levels are usually normal and intestinal absorption of calcium is normal or slightly reduced. Skeletal consequences of FHH are minor with the suggestion of a slightly increased bone turnover than normal being matched by increased bone formation.[169,170] Affected individuals from some FHH kindreds may experience pancreatitis, gallstones, or chondrocalcinosis.[166,167,171] The CASR is expressed at low levels in pancreas,[23] liver,[22] and chondrocytes[12,172] and decreased activity of the CASR could have a functional effect in these tissues.

The degree of hypercalcemia in most FHH patients is similar to that of patients with mild primary hyperparathyroidism (PHPT).[168] In contrast, serum magnesium levels are at the upper end of the normal range or slightly above.[173] The serum PTH levels are inappropriately normal (given the hypercalcemia) pointing to a derangement of the sensing of the blood calcium level by the parathyroid gland.[174,175] Diagnostic difficulties can arise in the ~15% of PHPT patients who have serum PTH levels at the upper limit of normal.[176] Because of this, FHH patients often underwent parathyroidectomy.[177–179] However, in the vast majority of the cases, the patients remained hypercalcemic because the defect in renal calcium sensing was uncorrected[180–182] and the consensus is that surgery should be avoided in this benign disorder.

There is unusually high renal tubular reabsorption of calcium and magnesium for the prevailing blood calcium concentration. In FHH patients, the renal calcium/creatinine clearance ratio is less than 1% while in PHPT and other hypercalcemic disorders it is usually much higher.[167] However, there is some overlap and some families with FHH have affected members with hypercalciuria and/or nephrolithiasis.[183–185] In FHH, unlike in PHPT, urinary concentrating ability is normal.[186] The renal CASR may mediate the polyuria and diminished urinary concentrating ability characteristic of the hypercalcemia in PHPT. One proposed mechanism is that in the renal IMCD the apically expressed CASR regulates (counteracts) vasopressin-controlled water permeability.[187]

B. Neonatal Severe Hyperparathyroidism

PHPT is extremely rare in childhood.[188,189] In reviewing several kindreds with FHH, it was noted that three patients had severe hyperparathyroidism in the neonatal period.[167] These and other cases involved multiglandular parathyroid hyperplasia rather than adenoma.[190–192] Neonatal severe hyperparathyroidism (NSHPT) occurs in children under the age of 6 months who have marked symptomatic hypercalcemia with the bony changes of hyperparathyroidism. In its most severe form, NSHPT can be a devastating neurodevelopmental disorder and is often fatal unless treated by total parathyroidectomy.[193,194] However, some cases (and in most instances not associated with FHH kindreds) present with hypercalcemia that is less marked and/or is transient (self-limited) and respond well to medical management.[195]

NSHPT can result in the following way: (1) The neonate is homozygous for the CASR mutation with each parent of a consanguineous union having one copy of the mutated allele and thus each having FHH.[194,196] (2) The neonate is a compound heterozygote carrying two different mutations each one coming from the individual parents.[197] (3) The neonate is heterozygous for the mutation and the mother is unaffected. This situation may arise either with a *de novo* mutation[198] or in the case of a paternal mutation.[81] Gestation of the FHH fetus in a normal mother will induce fetal secondary hyperparathyroidism because the fetal parathyroid glands will perceive the normal calcium level set by the mother as low.[199] These are the cases that are often self-limited.

Rare cases have been noted in which an infant (having a wild-type CASR genotype) and born of a mother with ADH may present neonatally with low-normal serum calcium and high-normal PTH levels. With a normal CASR, the fetal parathyroid glands respond *in utero* to the maternal hypocalcemia with an increased output of PTH.[200]

C. Molecular Genetics of FHH and NSHPT

In FHH, the inheritance is autosomal dominant with virtually 100% penetrance but variable expressivity. The disease locus maps in most cases to chromosome 3q (FHH type 1).[201–203] In four families with FHH and NSHPT, in which there was evidence of consanguinity, haplotype analysis with chromosome 3q markers linked to the FHH locus was consistent with NSHPT being the homozygous expression of the FHH disorder.[26] The *CASR* gene, itself, was shown by hamster–human hybrid cell hybridization analysis to reside on chromosome 3.[25] The *CASR* gene was mapped to 3q13.3–21 by fluorescence *in situ* hybridization (FISH) analysis.[34] However, the FHH trait exhibits genetic heterogeneity. In one family the disorder was mapped to 19p13.3 (FHH type 2)[202] and in another kindred with atypical features such as osteomalacia in some affected members and increased circulating PTH with age the condition mapped to 19q13 (FHH type 3).[204–206] At present, the genes responsible for the trait at the chromosome 19 loci are unknown.

D. Autosomal Dominant Hypocalcemia

Primary hypoparathyroidism encompasses a heterogeneous group of conditions in which hypocalcemia and hyperphosphatemia occur as a result of deficient PTH secretion. Familial isolated hypoparathyroidism (FIH; MIM# 146200) shows multiple modes of inheritance. Autosomal recessive FIH occurs with homozygous inactivating mutations of the PTH gene or the glial cells missing-2 (GCM2) gene that encodes a transcription factor expressed in the PTH-secreting cells of the parathyroid gland and is essential for their development in terrestrial vertebrates (41, see Ref. 207 and references therein). Autosomal dominant inheritance occurs with heterozygous inactivating mutations of the PTH gene (MIM# 168450)[208] or the GCM2 gene[41,209] that act in a dominant-negative fashion.

Hypoparathyroidism segregated in a large family as an autosomal dominant trait and was linked to chromosome 3q.[29] In this family, and others, heterozygous activating mutations of the *CASR* gene were identified.[28,30] The descriptor, ADH (MIM# 601198), rather than hypoparathyroidism, is now more commonly used for this familial hypocalcemic disorder.[210]

In ADH, the hypocalcemia can be mild to moderate and patients have relatively few symptoms. Seizures can occur, especially in younger patients, and these often occur during febrile episodes due to intercurrent infections. Parathesias, tetany, and laryngospasm occur but are uncommon. There is a tendency toward hyperphosphatemia, although serum phosphate levels may be normal. The renal CASR can function as a magnesium sensor[162,163] and serum magnesium levels can be at the lower end of normal or even below normal in untreated ADH patients. Serum intact PTH levels although low, are usually within the normal range. Urinary calcium excretion is higher than in hypoparathyroid patients of other etiologies, despite serum PTH levels being lower in the latter patient group.[31]

Renal tubular responsiveness to vitamin D and its metabolites is much more exuberant in ADH patients than in other hypoparathyroid patients. With the marked hypercalciuria, there is a greater risk of renal complications such as nephrocalcinosis, nephrolithiasis, and impaired renal function. The active metabolite of vitamin D, $1,25(OH)_2D$, enhances renal CASR expression.[39] Promotion of expression of the activated CASR excessively inhibits the reabsorption of calcium by the renal tubular cells sustaining the hypercalciuria. Therefore, ADH patients should be treated sparingly with vitamin D or calcium supplements.[31]

E. Bartter's Syndrome Subtype V

Although Bartter's syndrome subtype V (MIM# 146200) is represented by only a handful of cases due to heterozygous severe activating mutations in the CASR,[211–213] they provide special insight into the functioning of the CASR in the thick ascending limb of the nephron.[160] The Bartter syndrome encompasses a heterogeneous group of electrolyte homeostasis disorders, the common features of which are hypokalemic alkalosis, hyperreninemia, and hyperaldosteronism.[214] Bartter syndrome subtypes I–IV are autosomal recessive disorders due to inactivating mutations in the following ion transporters or channels active in the thick ascending limb of the loop of Henle; type I, the sodium–potassium–chloride cotransporter (NKCC2 or SLC12A2); type 2, the outwardly rectifying potassium channel (ROMK); type III, the voltage-gated chloride channel (CLC-Kb); type IV, Barttin, a β subunit that is required for trafficking of CLC-Ka and CLC-Kb. Patients with Bartter's syndrome subtype V have, in addition to the classic features of the syndrome, hypocalcemia and may exhibit neuromuscular manifestations, seizures, and basal ganglia calcifications.

In the cells of the renal thick ascending limb of the loop of Henle, the NKCC2 cotransporter is situated in the apical membrane (luminal side) and facilitates entry of Na^+, K^+, and Cl^-. Luminal K^+ is rate limiting for NKCC2 activity. The K^+ that enters the cell is recycled to the lumen via the apical potassium channel, ROMK. At the basolateral (blood side) membrane, exit of Na^+ occurs via the Na^+–K^+-ATPase and Cl^- exits via the voltage-gated chloride channel, ClC-Kb. ClC-Kb function requires interaction with the subunit Barttin. The transepithelial electrochemical gradient that is set up drives paracellular transport of Na^+, Mg^{2+}, and Ca^{2+} from the lumen to the blood. The CASR is situated in the basolateral membrane and when activated increases intracellular 20-HETE and decreases cAMP concentrations that inhibit ROMK and NKCC2 activities.[214] Thus, severe activating mutations of the CASR lead to the phenotype of Bartter's syndrome in addition to the hypocalcemic hypercalciuria of ADH (Table I).

XV. CASR Mutation Repertoire

Over 200 unique mutations in the CASR have been identified, 135 of the inactivating, FHH/NSHPT type, and 68 of the activating, ADH type (Fig. 6). There are 20 recurrent inactivating and 14 recurrent activating mutations found in apparently unrelated families. There are over 300 reports of families or individuals presenting independently with a CASR mutation (see http://www.casrdb.mcgill.ca).[215,216]

A. CASR-Inactivating Mutations (Table II)

The majority of the inactivating mutations are missense, single amino acid substitutions, and some insertion, deletion, frame shift, truncation, and splice-site mutations have been described. The mutations are scattered throughout the protein sequence with some clustering in the first half of the ECD (amino acids 137–250 within the VFT), the latter part of the ECD (amino acids 549–595 within the cysteine-rich region), and parts of the TMD. The scattering of mutations is consistent with the notion of the CASR having multiple functional components that collectively contribute to activity and that a critical mutation in any one of them can cause major impairment in function.[217,218]

TABLE I
Disorders Associated with the CASR

Familial (benign) hypocalciuric hypercalcemia (FHH)
- Mild hypercalcemia and hypermagnesemia
- PTH inappropriately normal
- Parathyroidectomy ineffective in normalizing hypercalcemia
- Renal calcium clearance ratio <0.01
- Renal concentrating ability normal

Neonatal severe hyperparathyroidism (NSHPT)
- Marked symptomatic hypercalcemia
- Bony changes of hyperparathyroidism
- Neurodevelopmental deficits if untreated
- Parathyroidectomy recommended
- Some less severe (self-limiting) forms

Autosomal dominant hypocalcemia (ADH)
- Mild to moderate hypocalcemia and hypomagnesemia
- Symptoms: seizures in younger patients
- Parathesias, tetany, and laryngospasm uncommon
- Urinary calcium excretion relatively increased
- Bartters subtype V in a few severe cases

Autoimmune hypocalciuric hypercalcemia (AHH)
- Phenocopy of FHH
- Inactivating CASR autoantibodies

Acquired hypoparathyroidism (AH)
- Phenocopy of ADH
- Activating CASR autoantibodies

Several mechanisms can account for impairment in CASR function: (1) Impairment in biosynthesis of the receptor and/or increased degradation.[44] (2) Normal biosynthesis and expression of the receptor within the cell but defective trafficking of the receptor from the ER to the plasma membrane.[92] (3) Normal maturation and trafficking of the receptor and expression on the plasma membrane but reduced affinity for calcium or other defect in the coupling of activation of the VFT to conformational changes in the TMD.[81] (4) Inability of the cell-surface CASR to couple normally to G proteins required for activation of cell signaling pathways. (5) An additional consideration applies

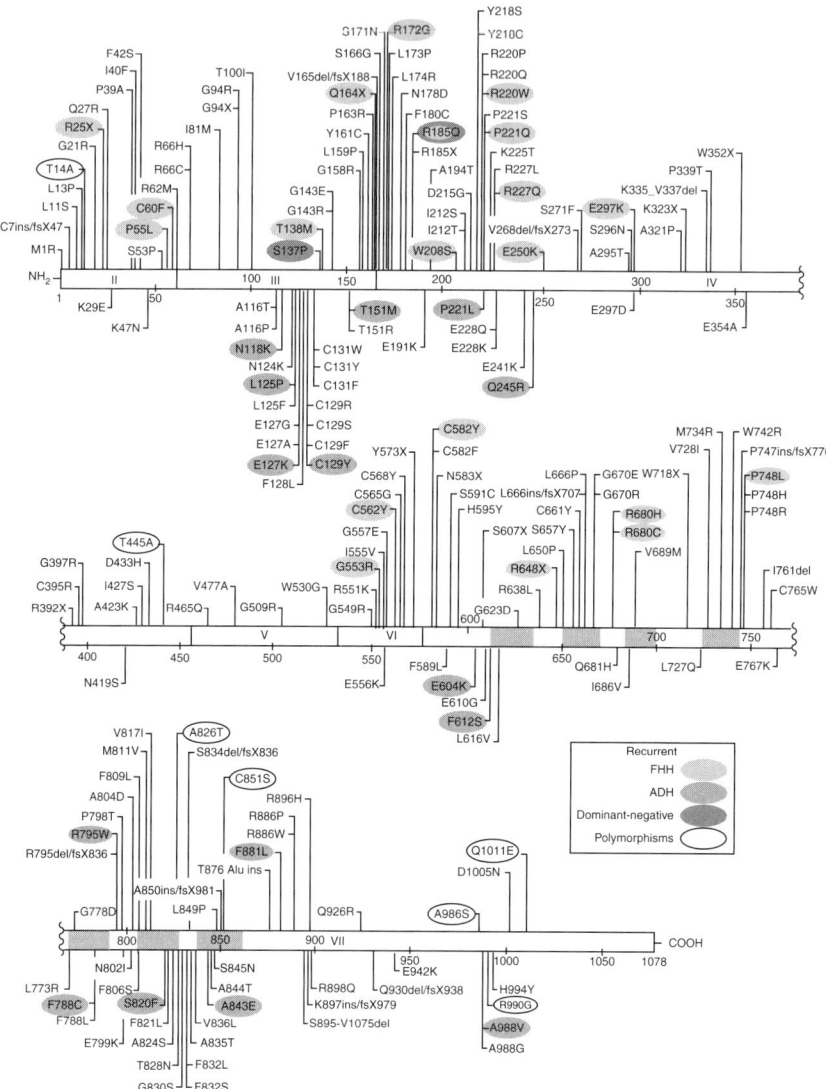

FIG. 6. Mutations and polymorphisms in the CASR. Schema of the CASR (which has 1078 amino acids) showing the relationship between gene exons (II–VII) and the portions of the protein they encode. Exons II–VI and the beginning of exon VII encode the ECD of ∼600 amino acids, exon VII encodes the TMD of ∼250 amino acids including membrane-spanning helices TM1–TM7 (indicated by the shaded boxes), ECL1–ECL3, ICL1–ICL3, and the ICD of ∼200 amino acids. The locations of the inactivating mutations found in patients with FHH and/or NSHPT (as well as some polymorphisms) are shown above and of the activating mutations found in patients with ADH (as well as polymorphism R990G) are shown below the bars representing the protein-coding exons.

to those mutants that by virtue of their ability to dimerize with the normal CASR encoded by the wild-type allele (in FHH), may exert a dominant-negative effect reducing the function of the wild-type CASR.[198,219,220]

One mutation, M1R, would disrupt the initiation codon altering the protein start site (**AXXATGG**), with three potential start sites upstream of the authentic site that would generate short peptides in a different reading frame and one downstream and in frame that would generate a protein beginning at amino acid 74. The precise consequences of the mutation have not been examined although expression levels are much reduced.[221]

The CASR has an NH_2-terminal 19-amino acid-signal peptide that directs the nascent polypeptide chain as it emerges from the ribosome, into the ER. Two mutations, L11S and L13P, lie within the signal peptide hydrophobic core.[44,222] When transiently transfected into kidney cells, the L11S and L13P mutants demonstrated markedly reduced intracellular and plasma membrane expression and signaling via the MAPK pathway in response to elevations in extracellular calcium relative to wild-type CASR. In cotranslational processing assays, which test the functionality of the signal peptide in the early secretory pathway, in contrast to the wild type, the mutants failed to be inserted into microsomes (representing the ER) and undergo core N-glycosylation. Hence, these mutants did not achieve proper biosynthesis and when coexpressed with the wild-type CASR did not influence the ability of the wild-type CASR to activate the MAPK signaling pathway at all.[44]

The mature CASR protein begins at amino acid 20. The three-dimensional globular lobes 1 and 2 of the VFT domain are formed by discontinuous protein sequence starting at amino acid 23 and ending at amino acid 528. A hinge region is formed by three strands of the sequence that weaves back and forth between the two lobes. Between amino acids 21–100, several inactivating missense mutations are present. Mutations S53P and the recurrent P55L occur in the unstructured loop 1 (amino acids 50–59) that when deleted leads to reduced receptor activity.[82] The C60F mutation would break an intramolecular disulfide bond with C101 critical for formation of the VFT structure.[90] The R66H and R66C mutations represent a class of mutant that undergoes dimerization as normal but does not exit the ER and never achieves localization at the plasma membrane.[92] From amino acids 115–131, inactivating mutations are absent but rather there is a very critical clustering of activating mutations in this region.

From amino acids 137–227, there are two dense clusters of inactivating missense mutations and then further mutations at a lower density extending to 339 and onwards. Many of these are likely to be involved (either directly or indirectly) in the putative ligand-binding domains that have been modeled. One study proposed a Ca^{2+}-binding site comprising polar residues—S170, D190, Q193, S296, and E297—directly involved in Ca^{2+}-binding with an additional set of residues contributing to a so-called coordination sphere of the cation (F270, Y218, S147).[93] The cluster of missense mutations from amino acids 158–185 as well as those from 271 onwards could contribute to alterations in the Ca^{2+}-binding site while other groupings of mutations (137–143, 208–227, and 252–352) could have an impact upon the coordination sphere of the cation. Another group has proposed that amino acids 132–300 be designated as a subdomain of the VFT having three putative Ca^{2+}-binding sites.[95] Site 1 is in the hinge region of the VFT and involves residues S147, S170, L174, R185, D190, Y218, P221, and E297. This would appear to overlap with the Ca^{2+}-binding site proposed by the previous study. Naturally occurring mutations of a selection of these residues (L174R, R185Q, Y218S, Y218C, P221Q, P221S, and E297K), some of which are recurrent, have been identified. The three adjacent serines, residues 169–171, have been proposed to be a binding site for aromatic amino acids such as phenylalanine and tyrosine that function as allosteric activators[223] and there is a naturally occurring S171N mutation. Ca^{2+}-binding site 2 involves residues L242, S244, Q245, D248, E250, and Q253. The recurrent inactivating mutation, E250K, has been reported. Site 3 involves residues R220, E228, E229, E231, and E232 and the recurrent inactivating mutation, R220W, has been described. For some of the residues within the Ca^{2+}-binding sites activating mutations have also been described and at some residues the resulting mutation can be either inactivating or activating depending upon the nature of the replacement amino acid.

Some of the missense mutations are located in the hinge region and could affect mobility of the hinge. The F180C mutation while expressed normally at the cell surface potentially can form a disulfide bond with the normally unlinked C482 residue thereby inhibiting closure of the lobes leading to markedly reduced cell signaling.[81] From amino acids 250–530, missense mutations are found but at much lower density than in the regions already discussed. The C395R mutation would disrupt a critical disulfide bond with C358 contributing to the VFT domain structure.[90] From amino acids 549–595, a dense clustering of mutations occurs within the critical cysteine-rich region. Several of these involve the cysteine residues themselves (C562Y, C565G, C568Y, C582Y, and C582F).

Many mutations are found within the TMD with several in α-helices TM2, TM3, and TM6. Although valuable information has come from modeling how the allosteric modifiers might interact with the TMD (see earlier section), to a large extent, it is still unclear how specific mutations within the TMD modify function. Further studies are needed in this area. With respect to mutations in the extracellular loops, there is C765W in ECL2 that would disrupt the critical disulfide bond formed with C677 in ECL1. Recurrent mutations of P748 at the junction of TM4 and ECL2 (P748H, P748L, and P748R) have been recorded. P748H was well expressed at the cell surface but exhibited reduced responsiveness to extracellular Ca^{2+}.[120] Several mutations are in ECL3 (and the extracellular part of TM6) potentially pointing to critical interaction of the ECD with this extracellular loop. Rather few inactivating mutations occur in the ICD, the F881L and R886P mutations were identified in cases of familial isolated hyperparathyroidism (FIHP) perhaps pointing to this part of the receptor being important in the parathyroid cell and less so in the nephron (Table II).

B. CASR-Activating Mutations (Table III)

Within the ECD, an important clustering of activating mutations is found from amino acids 116–131 in the loop of each protomer critical for intermolecular bonding to form the CASR dimer. Several mutations are found at C129 (C129F, C129R, C129S, C129Y) and C131 (C131F, C131Y, C131W). The intermolecular covalent linkages—129-S–S-129 and 131-S–S-131—contribute to the maintenance of the receptor in an inactivate state. Consistent with this notion, deletion of loop 2 (amino acids 114–126) in an engineered mutant led to increased sensitivity to Ca^{2+}.[82]

From amino acids 151–297, there is a scattering of activating mutations and these are within the Ca^{2+}-binding sites described above under inactivating mutations. The T151M mutation was found in a very large ADH kindred, one branch of which also had neoplasms such as pituitary adenoma, prostate carcinoma, or medulloblastoma, tissues that express the CASR.[224] Cadmium, an environmental carcinogen, was shown to transform NIH3T3 cells expressing the mutant but not wild-type CASR. This particular branch of the family came from a heavily industrialized area of Texas and it was postulated that the combination of toxic metal exposure and this particular CASR mutant stimulated oncogenesis in some family members.[224] E297D is in Ca^{2+}-binding site 1, Q245R is in site 2, and E228Q and E228K are in site 3. Of interest is that while P221L is activating, different substitutions at the same residue, P221S and P221Q are inactivating. Likewise, while E297D is activating, E297K is inactivating. For the changes at P221, an altered effect on the α-helical structure has been proposed.[95] For E297, 297D could promote enhanced Ca^{2+}-binding whereas the oppositely charged 297K could repulse Ca^{2+} binding.

Few other activating mutations occur in the remainder of the ECD but some mutations are present from amino acids 589–612 within the peptide linker region described earlier. Studies with engineered mutants had suggested this region is important for transmitting the activation signal from the ECD to the TMD. Within the TMD, the E767K in ECL2 mutation is of particular interest given that attention had been focused upon this residue by studies of engineered mutants with the change from a negatively charged amino acid to either a positive or neutral amino acid being activating. A few other activating mutations are scattered along the TMD but there is a striking clustering at ECL3 and the extracellular parts of TM6 and TM7. The recurrent A843E

TABLE II
Inactivating Mutations in the *CASR*

Region affected[a]	Mutation name[b]	Disease[c]	Author	No.
E2-5′UTR	c. − 10 acg>atg	FHH	Christie et al. (2007)	302
E2-SP	M1R	FHH/NSHPT	De Andrade et al. (2006)	221
E2-SP	C7ins/fsX47	FHH	D'Souza-Li et al. (2002)	218
E2-SP	L11S	FHH	Pidasheva et al. (2005)	44
E2-SP	L13P	FHH/NSHPT	Miyashiro et al. (2004)	222
E2-SP	T14A	Poly	Pidasheva et al. (2005)	44
E2-ECD	G21R	FHH	Nissen et al. (2007)	344
E2-ECD	R25X	NSHPT	Despert et al. (2005)	308
E2-ECD	R25X	FHH	Ward et al. (2006)	368
E2-ECD	R25X	NSHPT	Christie et al. (2007)	302
E2-ECD	Q27R	NSHPT	Chikatsu et al. (1999)	298
E2-ECD	P39A	FHH/NSHPT	Aida et al. (1995a)	292
E2-ECD	I40F	FHH	Vargas-Poussou et al. (2002)	211
E2-ECD	F42S	FHH	Cole et al. (2009)	225
E2-ECD	S53P	FHH	Heath et al. (1996)	217
E2-ECD	P55L	FHH	Pearce et al. (1995)	203
E2-ECD	P55L	FHH	Heath et al. (1996)	217
E2-ECD	P55L	FHH	Fukumoto et al. (2001)	316
E2-ECD	P55L	FHH	Cetani et al. (2003)	119
E2-ECD	P55L	FHH	Speer et al. (2003)	358
E2-ECD	P55L	FHH	Cole et al. (2009)	225
E2-ECD	C60F	NSHPT	Waller et al. (2004)	365
E2-ECD	C60F	NSHPT	Christie et al. (2007)	302
E2-ECD	R62M	FHH/NSHPT	Chou et al. (1995)	300
IVS2	+ 1g>c	FHH	Cetani et al. (2008)	120
IVS2	− 1g>t	FHH/NSHPT	D'Souza-Li et al. (2001)	310
E3-ECD	R66C	FHH/NSHPT	Chou et al. (1995)	300
E3-ECD	R66H	NSHPT	Pidasheva et al. (2006)	92
E3-ECD	I81M	FHH	Cole et al. (2009)	225
E3-ECD	G94R	FHH	Defrance-Faivre et al. (2008)	305
E3-ECD	G94X	NSHPT	Ward et al. (2004)	367
E3-ECD	T100I	FIHP	Warner et al. (2004)	269

(*Continues*)

TABLE II (Continued)

Region affected[a]	Mutation name[b]	Disease[c]	Author	No.
E3-ECD	S137P	FHH	Soie et al. (1999)	267
E3-ECD	T138M	FHH	Chou et al. (1995)	300
E3-ECD	T138M	FHH	D'Souza-Li et al. (2002)	218
E3-ECD	G143E	FHH	Chou et al. (1995)	300
E3-ECD	G143R	FHH	Cole et al. (2009)	225
E3-ECD	G158R	FHH	Cole et al. (2009)	225
E3-ECD	L159P	FIHP	Simonds et al. (2002)	268
E3-ECD	Y161C	NSHPT	Rajguru et al. (2001)	350
E3-ECD	P163R	TCP	Murugaian et al. (2008)	341
E3-ECD	Q164X	NSHPT	Waller et al. (2004)	365
E3-ECD	Q164X	NSHPT	Christie et al. (2007)	302
E3-ECD	Q164del/fsX188	NSHPT	Ward et al. (2006)	366
E3-ECD	S166G	FHH	Cole et al. (2009)	225
E4-ECD	S171N	FHH	Nissen et al. (2007)	344
E4-ECD	R172G	FHH	Christie et al. (2007)	302
E4-ECD	R172G	FHH	Henn et al. (2008)	320
E4-ECD	L173P	FHH	Felderbauer et al. (2005)	313
E4-ECD	L174R	FHH	Ward et al. (1997)	369
E4-ECD	N178D	FHH	Pearce et al. (1996)	184
E4-ECD	F180C	FHH	Zajickova et al. (2007)	81
E4-ECD	R185X	FHH/NSHPT	Kobayashi et al. (1997)	197
E4-ECD	R185Q	FHH	Pollak et al. (1993)	25
E4-ECD	R185Q	FHH	Heath et al. (1996)	217
E4-ECD	R185Q	NSHPT	Bai et al. (1997)	198
E4-ECD	R185Q	FHH	Sarli et al. (2004)	353
E4-ECD	R185Q	FHH	Christie et al. (2007)	302
E4-ECD	R185Q	NSHPT	Obermannova et al. (2009)	346
E4-ECD	A194T	FHH	Sinha et al. (2008)	357
E4-ECD	W208S	FHH	Christie et al. (2007)	302
E4-ECD	W208S	FHH	Demedts et al. (2008)	307
E4-ECD	I212S	NSHPT	Christie et al. (2007)	302
E4-ECD	I212T	FHH	Marcocci et al. (2003)	337
E4-ECD	D215G	FHH	Heath et al. (1996)	217

(Continues)

TABLE II (Continued)

Region affected[a]	Mutation name[b]	Disease[c]	Author	No.
E4-ECD	Y218C	FHH	Cetani et al. (2003)	119
E4-ECD	Y218S	FHH	Pearce et al. (1995)	203
E4-ECD	R220P	FHH	Ryan et al. (2005)	351
E4-ECD	R220W	FHH	Fukumoto et al. (2001)	316
E4-ECD	R220W	FHH	Schwarz et al. (2000)	355
E4-ECD	R220W	FHH	D'Souza-Li et al. (2002)	218
E4-ECD	R220W	FIHP	Simonds et al. (2002)	268
E4-ECD	R220W	FHH	Christie et al. (2007)	302
E4-ECD	R220W	NSHPT	Fox et al. (2007)	315
E4-ECD	R220W	FHH	Festen-Spanjer et al. (2007)	291
E4-ECD	R220W	FHH	Cole et al. (2009)	225
E4-ECD	R220Q	FHH	Pearce et al. (1996)	184
E4-ECD	P221S	FHH	Pearce et al. (1996)	184
E4-ECD	P221Q	FHH	Christie et al. (2007)	302
E4-ECD	P221Q	FHH	Nissen et al. (2007)	344
E4-ECD	K225T	FHH	Nissen et al. (2007)	344
E4-ECD	R227Q	FHH	Chou et al. (1995)	300
E4-ECD	R227Q	FHH	Wystrychowski et al. (2005)	220
E4-ECD	R227L	NSHPT	Pearce et al. (1995)	203
E4-ECD	E250K	FIHP	Simonds et al. (2002)	268
E4-ECD	E250K	FHH	Nissen et al. (2007)	344
E4-ECD	V268del/fsX273	FIHP	Simonds et al. (2002)	268
E4-ECD	S271F	FHH	Nissen et al. (2007)	344
E4-ECD	A295T	FHH	Defrance-Faivre et al. (2008)	305
E4-ECD	S296N	FHH	Christie et al. (2007)	302
E4-ECD	E297K	FHH/NSHPT	Pollak et al. (1993)	25
E4-ECD	E297K	FHH	Woo et al. (2006)	371
E4-ECD	A321P	FHH	Ono et al. (2008)	348
E4-ECD	K323X	FHH	Ward et al. (2006)	368
E4-ECD	K335_V337del	FIHP	Warner et al. (2004)	269
E4-ECD	P339T	FHH	Hannan et al. (2007)	318
E4-ECD	W352X	FHH	Nissen et al. (2007)	344
E4-ECD	R392X	NSHPT	Christie et al. (2007)	302

(Continues)

TABLE II (Continued)

Region affected[a]	Mutation name[b]	Disease[c]	References Author	No.
E4-ECD	C395R	FHH	Vigouroux et al. (2000)	364
E4-ECD	G397R	FHH	Nissen et al. (2007)	344
E4-ECD	A423K	FHH	Livadariu et al. (2008)	332
E4-ECD	I427S	TCP	Murugaian et al. (2008)	341
E4-ECD	D433H	TCP	Murugaian et al. (2008)	341
E4-ECD	T445A	Poly	Simonds et al. (2002)	268
E5-ECD	R465Q	FHH	Leech et al. (2006)	328
E5-ECD	V477A	TCP	Murugaian et al. (2008)	341
E5-ECD	G509R	FHH	Nissen et al. (2007)	344
E5-ECD	W530G	FHH	Rus et al. (2008)	123
E6-ECD	G549R	FHH	D'Souza-Li et al. (2002)	218
E6-ECD	R551K	NSHPT	Toke et al. (2007)	362
E6-ECD	G553R	FHH/NSHPT	Schwarz et al. (2000, family 1)	355
E6-ECD	G553R	FHH	Schwarz et al. (2000, family 2)	355
E6-ECD	G553R	FHH	Nissen et al. (2007, family 1)	344
E6-ECD	G553R	FHH	Nissen et al. (2007, family 2)	344
E6-ECD	I555V	FHH	Nissen et al. (2007)	344
E6-ECD	G557E	FHH	Nakayama et al. (2001)	343
E6-ECD	C562Y	FHH	Burski et al. (2002)	297
E6-ECD	C562Y	FHH	Nissen et al. (2007)	344
E6-ECD	C562Y	FHH	Nissen et al. (2007)	344
E6-ECD	C562Y	FHH	Cole et al. (2009)	225
E6-ECD	C565G	FHH	Cole et al. (2009)	225
E6-ECD	C568Y	FHH	Rus et al. (2008)	123
E6-ECD	Y573X	FHH	Nissen et al. (2007)	344
E7-ECD	C582Y	NSHPT	Pearce et al. (1995)	203
E7-ECD	C582Y	FHH	Nissen et al. (2007)	344
E7-ECD	C582Y	FHH	Cole et al. (2009)	225
E7-ECD	C582F	FHH	Nissen et al. (2007)	344
E7-ECD	N583X	FHH	Pidasheva et al. (2006)	92
E7-ECD	S591C	NSHPT	Nyweide et al. (2006)	345
E7-ECD	H595Y	FHH	Cetani et al. (2008)	120
E7-ECD	S607X	FHH	Pearce et al. (1995)	203

(Continues)

TABLE II (*Continued*)

Region affected[a]	Mutation name[b]	Disease[c]	Author	No.
E7-TM1	G623D	FHH	Nissen et al. (2007)	344
E7-ICL1	R638L	FHH/NSHPT	D'Souza-Li et al. (2002)	311
E7-ICL1	R648X	FHH	Jap et al. (2001)	325
E7-ICL1	R648X	FHH	Yamauchi et al. (2002)	373
E7-ICL1	R648X	FHH	Defrance-Faivre et al. (2008)	305
E7-ICL1	R648X	NSHPT	Ward et al. (2004)	367
E7-TM2	L650P	FIHP	Warner et al. (2004)	269
E7-TM2	S657Y	FHH	Heath et al. (1996)	217
E7-TM2	C661Y	FHH	Cole et al. (2009)	225
E7-TM2	L666ins/fsX707	FHH	Nissen et al. (2007)	344
E7-TM2	L666P	FHH	Defrance-Faivre et al. (2008)	305
E7-TM2	G670R	FHH	Pearce et al. (1995)	203
E7-TM2	G670E	FHH/NSHPT	Kobayashi et al. (1997)	197
E7-ECL1	R680C	FHH	Pearce et al. (1995)	203
E7-ECL1	R680C	NSHPT	Waller et al. (2004)	365
E7-ECL1	R680C	NSHPT	Christie et al. (2007)	302
E7-ECL1	R680H	FHH/NSHPT	Arunchaiya et al. (1998)	294
E7-ECL1	R680H	FHH	Cole et al. (2009)	225
E7-TM3	V689M	FIHP	Warner et al. (2004)	269
E7-ICL2	W718X	FHH	Rus et al. (2008)	123
E7-TM4	V728I	FHH	Nissen et al. (2007)	344
E7-TM4	M734R	FHH	Rus et al. (2008)	123
E7-TM4	W742R	FHH	Nissen et al. (2007)	344
E7-ECL2	P747ins/fsX776	NSHPT	Pearce et al. (1995)	203
E7-ECL2	P748H	FHH	Cetani et al. (2008)	120
E7-ECL2	P748L	FHH	Mahto et al. (2006)	335
E7-ECL2	P748L	FHH	Christie et al. (2007)	302
E7-ECL2	P748R	FHH	Heath et al. (1996)	217
E7-ECL2	I761del	FHH	Cole et al. (2009)	225
E7-ECL2	C765W	FHH	Cetani et al. (2008)	120
E7-TM5	G778D	FHH	Ward et al. (2006)	368
E7-ICL3	R795W	FHH	Pollak et al. (1993)	25
E7-ICL3	R795del/fsX836	FHH	Nissen et al. (2007)	344

(*Continues*)

TABLE II (Continued)

Region affected[a]	Mutation name[b]	Disease[c]	References Author	No.
E7-ICL3	P798T	FHH	Lam et al. (2005)	327
E7-ICL3	A804D	FHH/NSHPT	Miyashiro et al. (2004)	339
E7-TM6	F809L	FHH	Timmers et al. (2006)	290
E7-TM6	M811V	FHH	Cole et al. (2005)	303
E7-TM6	V817I	FHH	Pearce et al. (1995)	203
E7-TM6	A826T	Poly	Cetani et al. (1999)	272
E7-ECL3	S834del/fsX836	FHH	Ma et al. (2008)	334
E7-TM7	L849P	FHH	Rus et al. (2008)	123
E7-TM7	A850ins/fsX981	FHH	D'Souza-Li et al. (2002)	218
E7-TM7	C851S	Poly	Baron et al. (1996)	295
E7-TM7	C851S	Poly	Vigouroux et al. (2000)	364
E7-ICD	T876 Alu ins	FHH/NSHPT	Janicic et al. (1995)	196
E7-ICD	F881L	FHH	Carling et al. (2000)	185
E7-ICD	R886P	FIHP	Simonds et al. (2002)	268
E7-ICD	R886W	FHH	Nissen et al. (2007)	344
E7-ICD	R896H	FHH	Felderbauer et al. (2006)	314
E7-ICD	Q926R	FHH	Rus et al. (2008)	123
E7-ICD	A986S	Poly	Heath et al. (1996)	217
E7-ICD	D1005N	FHH	Rus et al. (2008)	123
E7-ICD	Q1011E	Poly	Heath et al. (1996)	217

[a]E1–E7, exons 1–7; IVS, intervening sequence; 5'-UTR, 5' untranslated region; SP, signal peptide; ECD, extracellular domain; TM1–7, transmembrane helices 1–7; ICL1–3, intracellular loops 1–3; ECL1–3, extracellular loops 1–3; ICD, intracellular domain.

[b]Mutation nomenclature according to Ref. 375.

[c]FHH, familial hypocalciuric hypercalcemia; NSHPT, neonatal severe hyperparathyroidism; FIHP, familial isolated hyperparathyroidism; TCP, tropical chronic pancreatitis; Poly, polymorphism.

mutation within TM7 is particularly active and has been found in patients with Bartter's syndrome subtype V. Further studies are needed to throw light on the mechanistic importance of this region.

Within the ICD, activating mutations are present in which almost all (S895-V1075del) or significant parts (K897ins/fsX979 and Q930del/fsX938) of the COOH-terminal tail are deleted. For the S895-V1075del mutant, increased cell-surface expression (relative to wild type) contributes to the activation. Missense mutations R898G, recurrent A988V, and A988G presented as idiopathic epilepsy and H994Y as idiopathic hypercalciuria rather than the typical picture of ADH. Thus, portions of the ICD may have selective importance in the role of the CASR in neural transmission or function of the distal nephron (Table III).

XVI. Autoantibodies and the CASR

CASR mutations are found in two-thirds of FHH kindreds. Even taking into account that the disorder is heterogeneous with a gene other than CASR being responsible in a few other kindreds, the etiology of one-third of cases is not known. The disease may be due to mutations in parts of the CASR gene not currently examined, for example, the promoters and introns and large insertions/deletions.[225] Some patients with anti-CASR autoantibodies (of the inactivating type) associated with autoimmune disorders such as sprue or autoimmune thyroid disease present as an FHH phenocopy termed acquired hypocalciuric hypercalcemia (AHH).[226–228] The anti-CASR antibodies are directed against the ECD and interfere with elevated extracellular Ca^{2+}-mediated suppression of PTH release and perturb Ca^{2+}-sensing in the kidney, thereby closely mimicking FHH.[226] The autoantibodies studied from one patient with AHH potentiated the calcium-activated G_q pathway leading to phosphatidyl inositide (PI) turnover while inhibiting the G_i pathway that phosphorylates ERK1/2.[228] The calcimimetic NPS-R-568, in the presence of extracellular calcium, overcame the effects of the antibody and increased both PI turnover and ERK1/2 activation. It was proposed that the antibody acted in an allosteric manner to maintain the CASR in an active conformation with respect to one signaling pathway only.

Autoantibodies from a subset of patients with autoimmune hypoparathyroidism that inhibited PTH secretion were identified several years ago.[229,230] More recently, the CASR has been identified as a self-antigen in patients with autoimmune polyendocrine syndrome type 1 (APS1) or acquired hypoparathyroidism associated with autoimmune hypothyroidism or idiopathic hypoparathyroidism.[231–235] The activating antibodies are directed against epitopes in the ECD of the receptor and increase IP_3 accumulation and activate MAPK in HEK293 cells stably expressing the CASR, and inhibit PTH secretion from parathyroid cells.[232]

XVII. CASR Polymorphisms

Various polymorphisms (both exonic and intronic) at the CASR locus have been identified. Linkage disequilibrium analysis of a large Caucasian cohort using the Haploview program showed that the CASR locus is divided into

TABLE III
Activating Mutations in the CASR

Region affected[a]	Mutation name[b]	Disease[c]	References Author	No.
E2-ECD	K29E	ADH	Hu et al. (2004)	323
E2-ECD	K47N	ADH	Okazaki et al. (1999)	347
E3-ECD	A116P	ADH	Christie et al. (2007)	302
E3-ECD	A116T	ADH	Baron et al. (1996)	295
E3-ECD	N118K	ADH	Pearce et al. (1996)	31
E3-ECD	N118K	ADH	De Luca et al. (1997)	306
E3-ECD	N118K	ADH	Cole et al. (2009)	225
E3-ECD	N124K	ADH	Hu et al. (2002)	322
E3-ECD	L125F	ADH	Cole et al. (2009)	225
E3-ECD	L125P	ADH	Sato et al. (2002)	354
E3-ECD	L125P	ADH	Vargas-Poussou et al. (2002)	211
E3-ECD	E127A	ADH	Pollak et al. (1994)	26
E3-ECD	E127G	ADH	Christie et al. (2007)	302
E3-ECD	E127K	ADH	Lienhardt et al. (2001)	32
E3-ECD	E127K	ADH	Christie et al. (2007)	302
E3-ECD	F128L	ADH	Pearce et al. (1996)	31
E3-ECD	C129F	ADH	Lienhardt et al. (2001)	32
E3-ECD	C129R	ADH	Cole et al. (2009)	225
E3-ECD	C129S	ADH	Hirai et al. (2001)	321
E3-ECD	C129Y	ADH	Burren et al. (2005)	296
E3-ECD	C129Y	ADH	Christie et al. (2007)	302
E3-ECD	C131W	ADH	Watanabe et al. (2002)	212
E3-ECD	C131F	ADH	Suzuki et al. (2005)	360
E3-ECD	C131Y	ADH	Christie et al. (2007)	302
E3-ECD	T151M	ADH	Lovlie et al. (1996)	333
E3-ECD	T151M	ADH	Pearce et al. (1996)	31
E3-ECD	T151M	ADH	Hoff et al. (1999)	224
E3-ECD	T151R	ADH	Haag et al. (2005)	317
E4-ECD	E191K	ADH	Pearce et al. (1996)	31
E4-ECD	P221L	ADH	Conley et al. (2000)	304
E4-ECD	P221L	ADH	Lienhardt et al. (2001)	32
E4-ECD	P221L	ADH	Poppe et al. (2002)	349

(Continues)

TABLE III (Continued)

Region affected[a]	Mutation name[b]	Disease[c]	Author	No.
E4-ECD	P221L	ADH	Chikatsu et al. (2003)	299
E4-ECD	P221L	ADH	Haag et al. (2005)	317
E4-ECD	P221L	ADH	Christie et al. (2007)	302
E4-ECD	E228K	ADH	Cole et al. (2009)	225
E4-ECD	E228Q	ADH	Conley et al. (2000)	304
E4-ECD	E241K	ADH	Christie et al. (2007)	302
E4-ECD	Q245R	ADH	Conley et al. (2000)	304
E4-ECD	E297D	ADH	Silve et al. (2005)	93
E4-ECD	E354A	IE	Kapoor et al. (2008)	326
E4-ECD	N419S	ADH	Christie et al. (2007)	302
E6-ECD	E556K	ADH	Livadariu et al. (2008)	332
E6-ECD	C565G	ADH	Cole et al. (2009)	225
E7-ECD	F589L	ADH	Leinhardt et al. (2001)	329
E7-ECD	E604K	ADH	Tan et al. (2003)	361
E7-ECD	E604K	ADH	Alvarez-Hernandez et al. (2003)	293
E7-ECD	E604K	ADH	Cole et al. (2009)	225
E7-ECD	E610G	ADH	Christie et al. (2007)	302
E7-ECD	F612S	ADH	Pearce et al. (1996)	31
E7-ECD	F612S	ADH	Mancilla et al. (1997)	336
E7-TM1	L616V	ADH	Stock et al. (1999)	359
E7-ECL1	Q681H	ADH	Baron et al. (1996)	295
E7-TM3	I686V	IE	Kapoor et al. (2008)	326
E7-TM4	L727Q	ADH	Mittelman et al. (2006)	338
E7-ECL2	E767K	ADH	Uckun-Kitapci et al. (2005)	363
E7-TM5	L773R	ADH	De Luca et al. (1997)	306
E7-TM5	F788C	ADH	Watanabe et al. (1998)	370
E7-TM5	F788C	ADH	Lienhardt et al. (2001)	32
E7-TM5	F788C	ADH	Mora et al. (2006)	340
E7-TM5	F788L	ADH	Hendy et al. (2003)	319
E7-ICL3	E799K	ADH	Lienhardt et al. (2001)	32
E7-ICL3	N802I	ADH	Cole et al. (2009)	225
E7-TM6	F806S	ADH	Baron et al. (1996)	295
E7-TM6	S820F	ADH	Yamamoto et al. (2000)	372

(Continues)

TABLE III (Continued)

Region affected[a]	Mutation name[b]	Disease[c]	References Author	No.
E7-TM6	F821L	ADH	Shiorara et al. (2004)	356
E7-TM6	A824S	ADH	Inoue et al. (1998)	324
E7-ECL3	T828N	ADH	Christie et al. (2007)	302
E7-ECL3	G830S	ADH	Cole et al. (2009)	225
E7-ECL3	F832L	ADH	Cole et al. (2009)	225
E7-ECL3	F832S	ADH	Dreimane et al. (2001)	309
E7-ECL3	A835T	ADH	D'Souza-Li et al. (2002)	218
E7-ECL3	V836L	ADH	Hu et al. (2005)	110
E7-TM7	A843E	ADH	Nakae et al. (1997)	342
E7-TM7	A843E	ADH	Zhao et al. (1999)	374
E7-TM7	A843E	ADH	Lienhardt et al. (2001)	32
E7-TM7	A843E	ADH	Watanabe et al. (2002)	212
E7-TM7	A843E	ADH	Sato et al. (2002)	354
E7-TM7	A843E	ADH	Sanda et al. (2008)	352
E7-TM7	A844T	ADH	Haag et al. (2005)	317
E7-TM7	S845N	ADH	Christie et al. (2007)	302
E7-ICD	S895_V1075del	ADH	Lienhardt et al. (2000)	330
E7-ICD	K897ins/fsX979	ADH	Lienhardt et al. (2001)	331
E7-ICD	R898Q	IE	Kapoor et al. (2008)	326
E7-ICD	Q930del/fsX938	ADH	Christie et al. (2007)	302
E7-ICD	E942K	ADH	Ekhzaimy et al. (2006)	312
E7-ICD	A988V	IE	Kapoor et al. (2008, Case 1)	326
E7-ICD	A988V	IE	Kapoor et al. (2008, Case 2)	326
E7-ICD	A988G	IE	Kapoor et al. (2008)	326
E7-ICD	R990G	Poly	Heath et al. (1996)	217
E7-ICD	H994Y	IH	Christie et al. (2002)	301

[a]E1–E7, exons 1–7; ECD, extracellular domain; TM1–7, transmembrane helices 1–7; ICL1–3, intracellular loops 1–3; ECL1–3, extracellular loops 1–3; ICD, intracellular domain.
[b]Mutation nomenclature according to Ref. 375.
[c]ADH, autosomal dominant hypocalcemia; IE, idiopathic epilepsy; IH, idiopathic hypercalciuria; Poly, polymorphism.

(at least) three haplotype blocks, coincident with 5′ regulatory, coding, and 3′ domains.[35] Three single nucleotide polymorphisms (SNPs) in exon 7 encode nonconservative amino acid changes in the carboxy-terminal tail of the CASR protein.[217] The most common missense SNP in Caucasians, c.2956G > T, results in a serine substitution for alanine-986. The other nearby SNPs, c.2968A > G and c.3031C > G, encode a glycine substitution at arginine-990 and a glutamate substitution at glutamine-1011, respectively. Their role in disease pathogenesis is the subject of ongoing investigation. There are large ethnic differences in their frequencies[35] and linkage disequilibrium exists between them, making it difficult to isolate the effects of a single SNP. In some association studies, trilocus haplotypes may be better predictors than any single SNP genotype alone.[236,237]

Functionally activating and inactivating CASR mutations exert their most obvious clinical effects on serum and urine calcium, and it is these quantitative traits that have been scrutinized most with respect to CASR SNPs. Initial surveys suggested that both total and ionized serum calcium concentrations were increased (within the normal range) in young women carrying the 986S allele.[238–240] With phase ascertainment across the three missense SNPs, the trilocus haplotype was also predictive of blood ionized calcium in a cohort of healthy Italian men and women.[236] Similar significant differences[241,242] or (nonsignificant) trends[243–246] have been reported in healthy adult controls. On the other hand, several negative studies with apparently adequate design and sufficient power[247–249] suggest that ascertainment issues, such as founder effects or age of population or environmental confounders (e.g., vitamin D status) make replication difficult. Although a few studies have found associations between CASR polymorphisms and bone mineral density (BMD),[240,250,251] the majority have found no evidence for this.[246,249,252–255]

Some studies indicate a small but significant overall association between the PHPT phenotype and both A986S and R990G loci.[237,243,244] The minor 986S allele appears to act as a mild inactivating variant promoting increases in serum calcium and relative decreases in calcium excretion. A986S is not a strong predictor of hypercalciuria.[256] While 986S is more common in PHPT patients overall,[237] in those patients having renal stones it is less common with 990G being more frequent.[237,257] The 990G variant has been implicated in increased calcium excretion in idiopathic hypercalciuria consistent with it having an activating function in the kidney.[258,259]

In keeping with the widespread expression of the CASR, there is evidence that the three variants (A986S, R990G, or Q1011E) and other SNPs at the CASR locus may be associated with a more diverse set of phenotypes than just those associated with bone and mineral disorders. Association studies that have examined either single CASR variants or haplotypes as predictors for a variety of common disorders, including hypertension,[260,261] coronary artery disease,[241] cancer,[262–264] chronic pancreatitis,[265] and Alzheimer's disease[266] have provided evidence of a role of the CASR gene in these disorders.

XVIII. Altered Expression of CASR and Disease

Altered expression of the *CASR* may play a part in disease pathogenesis or progression (hyperparathyroidism, cancers), a resetting of the calciostat during inflammatory responses (septic shock, burn injury) or hyperresponsiveness of the renal CASR to vitamin D (in ADH patients).

A. Hyperparathyroidism

Hyperparathyroidism is usually not part of the FHH syndrome and histologically the parathyroid glands removed from FHH patients are only occasionally hyperplastic.[178,179] This contrasts with the marked hypercellularity of the glands removed from NSHPT patients and the *Casr* knockout mouse model.[149] However, some FHH kindreds (and FIHP kindreds with *CASR* mutations) have affected family members with elevated PTH levels and parathyroid tumors.[185,267–269] Thus, it may be that some mutant CASRs, even in the heterozygous state, have deleterious effects with respect to parathyroid cell proliferation. The link between parathyroid calcium sensing and proliferative pathways suggested that somatic alterations in the *CASR* gene could be tumorigenic. However, somatic mutation of the *CASR* gene rarely if ever contributes to the pathogenesis of sporadic parathyroid tumors.[270–273] On the other hand, the majority of the parathyroid glands of patients with primary and severe uremic secondary hyperparathyroidism have reduced *CASR* expression.[38,273,274] Thus, mutations in other genes regulating the calcium-sensing pathway may play a significant role in the initiation or the progression of parathyroid tumorigenesis.

Vitamin D insufficiency is common among patients with PHPT and vitamin D deficiency is associated with more severe and progressive disease. The active metabolite, $1,25(OH)_2D$, regulates parathyroid cell proliferation[275] and expression of the *PTH* and *CASR* genes.[39] Humans or mice in which there is homozygous inactivation of the VDR or its ligand[152] manifest marked parathyroid hyperplasia and elevated circulating PTH levels. Like the *CASR* gene, somatic mutation of the *VDR* gene does not contribute to parathyroid tumorigenesis, but VDR expression is reduced in both primary and secondary hyperparathyroid patients. Thus the reduced CASR expression may, in part, be secondary to decreased VDR expression. In FHH-affected individuals, the clinical severity can be exacerbated by vitamin D deficiency promoting overt hyperparathyroidism and vitamin D supplementation can restore the clinical picture to that of modestly elevated serum calcium with PTH levels within the normal range.[81]

B. Hypercalciuria

In the CTAL of the distal nephron, the activated CASR signals to inhibit the K^+ channel that drives the paracellular uptake of cations (see earlier section). Hence, the activated CASR promotes hypercalciuria. Patients with ADH provide a special management problem relative to other forms of hypoparathyroidism. The normal treatment with vitamin D metabolites carries the risk of excessively stimulating renal calcium excretion (while failing to bring serum calcium to the normal range) leading to nephrocalcinosis, nephrolithiasis, and renal damage. Upregulation by 1α-hydroxylated vitamin D metabolites exacerbates the responsiveness to cations by the already oversensitive CASR.[31,32]

Altered regulation of CASR expression by vitamin D metabolites may be critical in genetic hypercalciuria contributing to stone formation. A genetic hypercalciuric stone-forming rat model demonstrates features of human hypercalciuric nephrolithiasis. The rat model exhibits elevated levels of the VDR in the nephron and concomitantly increased CASR expression and defective renal calcium reabsorption.[276,277]

C. CASR and Cancer

The CASR is involved in normal cell proliferation and differentiation control in several tissues (not only the parathyroid gland). Alterations of CASR expression have been implicated in parathyroid neoplasia and breast, prostate and colon cancers.[278,279]

The CASR is expressed in human colon epithelium and regulates cell proliferation and differentiation. Cells of the colon crypt acquire CASR expression as they differentiate and migrate toward the apex of the crypt. CASR expression is weak or absent in colon carcinomas and is inversely correlated with differentiation status. Extracellular calcium and $1,25(OH)_2D$ upregulate CASR and cyclin-dependent kinase inhibitor, p21 and p27, expression in the colon and the chemopreventive activities of calcium and $1,25(OH)_2D$ in colon cancer may be mediated, in part, through the CASR.[280]

The CASR is expressed in both normal and malignant breast tissue and elevated levels are found in highly metastatic primary breast cancer cells and cell lines. In normal breast cells, activation of the CASR inhibits the release of the growth factor and promoter of bone metastases, parathyroid hormone-related peptide (PTHrP). In malignant breast cells, the CASR acts as an oncogene and stimulates the production of PTHrP.[48] The resulting hypercalcemia provides a potent stimulus to further activation of the breast CASR and release of PTHrP generating a so-called vicious cycle.[281] In a similar fashion, enhanced CASR expression and altered proliferation occur in prostate cancer cells.[282,283]

D. Proinflammatory Cytokines

Critically ill patients, with sepsis or burn injury, commonly have hypocalcemia.[284,285] The degree of hypocalcemia is inversely correlated with survival rate. The levels of circulating proinflammatory cytokines, interleukin-1β (IL-1β) and interleukin-6 (IL-6) are elevated and transactivate the CASR gene in parathyroid gland, thyroid C cells, and kidney.[40,42] PTH and 1,25 (OH)$_2$D levels are reduced and those of calcitonin increased contributing to the hypocalcemia. The mechanism may represent a critical counter-regulatory system that attempts to minimize the effects of calcium and cytokines in promoting intravascular coagulation and artherosclerosis during the inflammatory response.

XIX. CASR Allosteric Modifiers in the Clinic

Calcimimetics have been approved for use in patients with chronic kidney disease on dialysis and those with parathyroid carcinoma, and by their direct action on the parathyroid gland CASR they provide an effective medical means of lowering PTH secretion.[286,287] Cinacalcet HCl is marketed as Sensipar in North America and Australia and Mimpara in the European Union. Ongoing clinical trials in patients with mild PHPT have shown that calcimimetics reduce serum calcium and PTH levels and increase serum phosphate levels but do not significantly affect bone turnover or BMD.[288] While calcimimetics provide an important addition to the armamentarium of drugs to treat the secondary hyperparathyroidism of chronic kidney disease, their more widespread use in the medical management of PHPT is uncertain at present.

CASR antagonists, calcilytics, are also being evaluated in clinical trials as a treatment of osteoporosis. As intermittent exogenous administration of PTH produces increases in BMD, it is proposed that once-daily administration of a short-acting calcilytic could achieve a similar result by producing a pulse of endogenous PTH secretion (see Ref. 289 for review).

Other potential areas where the use of CASR allosteric modifiers are being or could be explored include (but are not restricted to) individual cases of FHH[290,291] or ADH, breast, prostate and other cancers, and the exuberant hypocalcemic response of critically ill patients.

XX. Summary

The central role of the parathyroid and kidney CASR as the calciostat controlling systemic mineral ion homeostasis is exemplified by the identification of inactivating and activating mutations in the gene as well as inactivating and activating autoantibodies. The widespread expression of the CASR predicts functions other than those involved in calcium homeostasis; and the calcimimetics and calilytics either in current or projected clinical use in mineral ion disorders may ultimately find application in a much broader array of diseases.

ACKNOWLEDGMENTS

Work from the authors' laboratory was supported by Canadian Institutes of Health Research (CIHR) Grants MOP-86581 and MOP-57730 (to G.N.H.). V.G. was supported by the Italian Health Ministry (Ricerca Oncologica 2006). L.C. was the recipient of a biomedical fellowship from the Kidney Foundation of Canada and a fellowship from the McGill University Health Center Research Institute.

REFERENCES

1. Brown EM, Gamba G, Riccardi D, Lombardi D, Butters RR, Kifor O, et al. Cloning and characterization of an extracellular Ca^{2+}-sensing receptor from bovine parathyroid. *Nature* 1993;**366**:575–80.
2. Brown EM. Four parameter model of the sigmoidal relationship between parathyroid hormone release and extracellular calcium concentration in normal and abnormal parathyroid tissue. *J Clin Endocrinol Metab* 1983;**56**:572–81.
3. Garrett JE, Tamir H, Kifor O, Simin RT, Rogers KV, Mithal A, et al. Calcitonin-secreting cells of the thyroid express an extracellular calcium receptor gene. *Endocrinology* 1995;**136**:5202–11.
4. Friechel M, Zink-Lorentz A, Holloschi A, Hafner M, Flocerzi V, Raue F. Expression of a calcium-sensing receptor in a human medullary cell line and its contribution to calcitonin secretion. *Endocrinology* 1996;**137**:3842–8.
5. Fudge NJ, Kovacs CS. Physiological studies in heterozygous calcium-sensing receptor (CaSR) gene-ablated mice confirm that CaSR regulates calcitonin release *in vivo*. *BMC Physiol* 2004;**4**:5.
6. VanHouten JN, Neville MC, Wysolmerski JJ. The calcium-sensing receptor regulates plasma membrane calcium adenosine triphosphatase isoform 2 activity in mammary epithelial cells: mechanism for calcium-regulated calcium transport into milk. *Endocrinology* 2002;**148**:5943–54.
7. Brown EM, Lian JB. New insights in bone biology: unmasking skeletal effects of the extracellular calcium-sensing receptor. *Sci Signal* 2008;**1**:pe40.
8. Chang W, Tu C, Chen TH, Bikle D, Shoback D. The extracellular calcium-sensing receptor (CaSR) is a critical modulator of skeletal development. *Sci Signal* 2008;**1**:ra1.
9. Yamaguchi T. The calcium-sensing receptor in bone. *J Bone Miner Metab* 2008;**26**:301–11.
10. Mentaverri R, Yano S, Chattopadhyay N, Petit L, Kifor O, Kamel S, et al. The calcium-sensing receptor is directly involved in both osteoclast differentiation and apoptosis. *FASEB J* 2006;**20**:2562–4.
11. Cheng L, Wang D, Palmer G, Caverzasio J, Canaff L, Hendy GN, et al. Regulation of expression of the calcium-sensing receptor (CASR) and sodium-dependent phosphate transporter (Glvr-1) in differentiating chondrocytes *in vivo*. *Bone* 1998;**23S**:SA114.

12. Chang W, Tu C, Chen TH, Komuves L, Oda Y, Pratt SA, et al. Expression and signal transduction of calcium-sensing receptors in cartilage and bone. *Endocrinology* 1999;**140**:5883–93.
13. Dvorak MM, Chen TH, Orwoll B, Garvey C, Chang W, Bikle DD, et al. Constitutive activity of the osteoblast Ca^{2+}-sensing receptor promotes loss of cancellous bone. *Endocrinology* 2007;**148**:3156–63.
14. Kovacs CS, Ho-Pao CL, Hunzelman JL, Lanske B, Fox J, Seidman JG, et al. Regulation of murine fetal-placental calcium metabolism by the calcium-sensing receptor. *J Clin Invest* 1998;**101**:2812–20.
15. Ruat M, Molliver ME, Snowman AM, Snyder SH. Calcium sensing receptor: molecular cloning in rat and localization to nerve terminals. *Proc Natl Acad Sci USA* 1995;**92**:3161–5.
16. Oda Y, Tu CL, Pillai S, Bikle D. The calcium-sensing receptor and its alternatively spliced form in keratinocyte differentiation. *J Biol Chem* 1998;**273**:23344–52.
17. Molostvov G, James S, Fletcher S, Bennett J, Lehnert H, Bland R, et al. Extracellular calcium-sensing receptor is functionally expressed in human artery. *Am J Physiol Renal Physiol* 2007;**293**:F946–55.
18. Adams GB, Chabner KT, Alley IR, Olson DP, Szczepiorkowski ZM, Poznansky MC, et al. Stem cell engraftment at the endosteal niche is specified by the calcium-sensing receptor. *Nature* 2006;**439**:599–603.
19. Ray JM, Squires PE, Curtis SB, Meloche MR, Buchan AM. Expression of the calcium-sensing receptor on human antral gastrin cells in culture. *J Clin Invest* 1997;**99**:2328–33.
20. Geibel J, Sritharan K, Geibel R, Persing JS, Seeher A, Roepke TK, et al. Calcium-sensing receptor abrogates secretagogue-induced increases in intestinal net fluid secretion by enhancing cyclic nucleotide destruction. *Proc Natl Acad Sci USA* 2006;**103**:9390–7.
21. Cheng SX, Okuda M, Hall AE, Geibel JP, Hebert SC. Expression of calcium-sensing receptor in rat colonic epithelium: evidence for modulation of fluid secretion. *Am J Physiol Gastrointest Liver Physiol* 2001;**283**:G240–50.
22. Canaff L, Petit J-L, Kisiel M, Watson PH, Gascon-Barré M, Hendy GN. Extracellular calcium-sensing receptor is expressed in rat hepatocytes: coupling to intracellular calcium mobilization and stimulation of bile flow. *J Biol Chem* 2001;**276**:4070–9.
23. Bruce JL, Xang XL, Ferguson CJ, Elliott AC, Steward MC, Case RM, et al. Molecular and functional identification of a Ca^{2+} (polyvalent cation)-sensing receptor in rat pancreas. *J Biol Chem* 1999;**274**:20561–8.
24. Brown EM, MacLeod RJ. Extracellular calcium sensing and extracellular calcium signaling. *Physiol Rev* 2001;**81**:239–97.
25. Pollak MR, Brown EM, Chou YHW, Hebert SC, Marx SJ, Steinmann B, et al. Mutations in the human Ca^{2+}-sensing receptor gene cause familial hypocalciuric hypercalcemia and neonatal severe hyperparathyroidism. *Cell* 1993;**75**:1297–303.
26. Pollak MR, Chou YH, Marx SJ, Steinmann B, Cole DEC, Brandi M, et al. Familial hypocalciuric hypercalcemia and neonatal severe hyperparathyroidism. Effects of mutant gene dosage on phenotype. *J Clin Invest* 1994;**93**:1108–12.
27. Brown EM. Clinical lessons from the calcium-sensing receptor. *Nat Clin Pract Endocrinol Metab* 2007;**3**:122–33.
28. Pollak MR, Brown EM, Estep HL, McLaine PN, Kifor O, Park J, et al. Autosomal dominant hypocalcemia caused by a Ca^{2+}-sensing receptor gene mutation. *Nat Genet* 1994;**8**:303–7.
29. Finegold DN, Armitage MM, Galiani M, Matise TC, Pandian MR, Perry YM, et al. Preliminary localization of a gene for autosomal dominant hypoparathyroidism to chromosome 3q13. *J Pediatr Res* 1994;**36**:414–7.
30. Perry YM, Finegold DN, Armitage MM, Ferrell RE. Missense mutation in the Ca-sensing receptor gene causes familial autosomal dominant hypoparathyroidism. *Am J Hum Genet* 1994; (Suppl. 55), A17, p 79.

31. Pearce SHS, Williamson C, Kifor O, Bai M, Coulthard MG, Davies M, et al. A familial syndrome of hypocalcemia with hypercalciuria due to mutations in the calcium-sensing receptor. *N Engl J Med* 1996;**335**:1115–22.
32. Lienhardt A, Bai M, Lagarde J-P, Rigaude M, Zhang Z, Jiang Y, et al. Activating mutations of the calcium-sensing receptor: management of hypocalcemia. *J Clin Endocrinol Metab* 2001;**86**:5313–23.
33. Brauner-Osborne H, Wellendorph P, Jensen AA. Structure, pharmacology and therapeutic prospects of family C G-protein coupled receptors. *Curr Drug Targets* 2007;**8**:169–84.
34. Janicic N, Soliman E, Pausova Z, Seldin MF, Riviere M, Szpirer J, et al. Mapping of the calcium-sensing receptor gene (CASR) to human chromosome 3q13.3–21 by fluorescence *in situ* hybridization, and localization to rat chromosome 11 and mouse chromosome16. *Mamm Genome* 1995;**6**:798–801.
35. Yun FHJ, Wong BYL, Chase M, Shuen AY, Canaff L, Thongthai K, et al. Genetic variation at the calcium-sensing receptor (*CASR*) locus: implications for clinical molecular diagnostics. *Clin Biochem* 2007;**40**:551–61.
36. Garrett JE, Capuano IV, Hammerland LG, Hung BC, Brown EM, Hebert SC, et al. Molecular cloning and functional expression of human parathyroid calcium receptor cDNAs. *J Biol Chem* 1995;**270**:12919–25.
37. Aida K, Koishi S, Tawata M, Onaya T. Molecular cloning of a putative Ca^{2+}-sensing receptor cDNA from human kidney. *Biochem Biophys Res Commun* 1995;**214**:524–9.
38. Chikatsu N, Fukumoto S, Takeuchi Y, Suzawa M, Obara T, Matsumoto T, et al. Cloning and characterization of two promoters for the human calcium-sensing receptor (CaSR) and changes of CaSR expression in parathyroid adenomas. *J Biol Chem* 2000;**275**:7553–7.
39. Canaff L, Hendy GN. Human calcium-sensing receptor gene: vitamin D response elements in promoters P1 and P2 confer transcriptional responsiveness to 1, 25-dihydroxyvitamin D. *J Biol Chem* 2002;**277**:30337–50.
40. Canaff L, Hendy GN. Calcium-sensing receptor gene transcription is upregulated by the proinflammatory cytokine, interleukin-1beta: role of the NF-kappaB pathway and kappaB elements. *J Biol Chem* 2005;**280**:14177–88.
41. Canaff L, Zhou X, Mosesova I, Cole DEC, Hendy GN. Glial cells missing-2 transactivates the calcium-sensing receptor gene: effect of a dominant-negative GCM2 mutant associated with autosomal dominant hypoparathyroidism. *Hum Mutat* 2009;**30**:85–92.
42. Canaff L, Zhou X, Hendy GN. The proinflammatory cytokine, interleukin-6, upregulates calcium-sensing receptor gene transcription via Stat1/3 and Sp1/3. *J Biol Chem* 2008;**283**:13586–600.
43. Goldsmith PK, Fan GF, Ray K, Shiloach J, McPhie P, Rogers KV, et al. Expression, purification, and biochemical characterization of the amino-terminal extracellular domain of the human calcium receptor. *J Biol Chem* 1999;**274**:11303–9.
44. Pidasheva S, Canaff L, Simonds WF, Marx SJ, Hendy GN. Impaired cotranslational processing of the calcium-sensing receptor due to signal peptide missense mutations in familial hypocalciuric hypercalcemia. *Hum Mol Genet* 2005;**14**:1679–90.
45. Hu J, Spiegel AM. Structure and function of the human calcium-sensing receptor: insights from natural and engineered mutations and allosteric modulators. *J Cell Mol Med* 2007;**11**:908–22.
46. Bai M. Structure–function relationship of the extracellular calcium-sensing receptor. *Cell Calcium* 2004;**35**:197–207.
47. Emanuel RL, Adler GK, Kifor O, Quinn SJ, Fuller F, Krapcho K, et al. Calcium-sensing receptor expression and regulation by extracellular calcium in the AtT-20 pituitary cell line. *Mol Endocrinol* 1996;**10**:555–65.

48. Mamillapalli R, VanHouten J, Zawalich W, Wysolmerski J. Switching of G-protein usage by the calcium-sensing receptor reverses its effect on parathyroid hormone-related protein secretion in normal versus malignant breast cells. *J Biol Chem* 2008;**283**:24435–47.
49. Ward DT. Calcium receptor-mediated intracellular signaling. *Cell Calcium* 2004;**35**:217–28.
50. Quinn SJ, Ye CP, Diaz R, Kifor O, Bai M, Vassilev P, et al. The Ca^{2+}-sensing receptor: a target for polyamines. *Am J Physiol* 1997;**273**:C1315–23.
51. Ye C, Ho-Pao CL, Kanazirska M, Quinn SJ, Rogers K, Seidman CE, et al. Amyloid-beta proteins activate Ca(2+)-permeable channels through calcium-sensing receptors. *J Neurosci Res* 1997;**47**:547–54.
52. Ward DT, Maldonado-Perez D, Hollins L, Riccardi D. Aminoglycosides induce acute cell signaling and chronic cell death in renal cells that express the calcium-sensing receptor. *J Am Soc Nephrol* 2005;**16**:1236–44.
53. Brauner-Osborne H, Jensen AA, Sheppard PO, O'Hara P, Krogsgaard-Larsen P. The agonist binding domain of the calcium-sensing receptor is located at the amino-terminal domain. *J Biol Chem* 1999;**274**:18382–6.
54. Quinn SJ, Kifor O, Trivedi S, Diaz R, Vassilev P, Brown EM. Sodium and ionic strength sensing by the calcium receptor. *J Biol Chem* 1998;**273**:19579–86.
55. Quinn SJ, Bai M, Brown EM. pH sensing by the calcium-sensing receptor. *J Biol Chem* 2004;**279**:37241–9.
56. Conigrave AD, Quinn SJ, Brown EM. L-Amino acid sensing by the extracellular Ca^{2+}-sensing receptor. *Proc Natl Acad Sci USA* 2000;**97**:4814–9.
57. Conigrave AD, Mun HC, Delbridge L, Quinn SJ, Wilkinson M, Brown EM. L-Amino acids regulate parathyroid hormone secretion. *J Biol Chem* 2004;**279**:38151–9.
58. Wellendorph P, Brauner-Osborne H. Molecular basis for amino acid sensing by family C G-protein-coupled receptors. *Brit J Pharmacol* 2009;**156**:869–84.
59. Conigrave AD, Brown EM, Rizzoli R. Dietary protein and bone health: roles of amino acid-sensing receptors in the control of calcium metabolism and bone homeostasis. *Annu Rev Nutr* 2008;**28**:131–55.
60. Nemeth EF, Fox J. Calcimimetic compounds: a direct approach to controlling plasma levels of parathyroid hormone in hyperparathyroidism. *Trends Endocrinol Metab* 1999;**10**:66–71.
61. Jensen AA, Brauner-Osborne H. Allosteric modulation of the calcium-sensing receptor. *Curr Neuropharmacol* 2007;**5**:180–6.
62. Hammerland LG, Garrett JE, Hung BCP, Levinthal C, Nemeth EF. Allosteric activation of the Ca^{2+} receptor expressed in *Xenopus laevis* oocytes by NPS 467 or NPS 568. *Mol Pharmacol* 1998;**53**:1083–8.
63. Nemeth EF, Steffey ME, Hammerland LG, Hung BCP, Van Wagenen BC, DelMar EG, et al. Calcimimetics with potent and selective activity on the parathyroid calcium receptor. *Proc Natl Acad Sci USA* 1998;**95**:4040–5.
64. Nemeth EF, Heaton WH, Miller M, Fox J, Balandrin MF, Van Wagenen BC, et al. Pharmacodynamics of the type II calcimimetic compound cinacalcet HCl. *J Pharmacol Exp Ther* 2004;**308**:627–35.
65. Dauban P, Ferry S, Faure H, Ruat M, Dodd RH. N1-Arylsulfonyl-N2-(1-aryl)ethyl-3-phenyl-propane-1, 2-diamines as novel calcimimetics acting on the calcium sensing receptor. *Bioorg Med Chem Lett* 2000;**10**:2001–4.
66. Kessler A, Faure H, Petrel C, Ruat M, Dauban P, Dodd RH. N2-Benzyl-N1-[1-(1-naphthyl)ethyl]-3-phenylpropane-1, 2-diamines and conformationally restrained indole analogues: development of calindol as a new calcimimetic acting at the calcium sensing receptor. *Bioorg Med Chem Lett* 2004;**14**:3345–9.
67. Nemeth EF, Delmar EG, Heaton WL, Miller MA, Lambert LD, Conklin RL, et al. Calcilytic compounds: potent and selective Ca^{2+} receptor antagonists that stimulate secretion of parathyroid hormone. *J Pharmacol Exp Ther* 2001;**299**:323–31.

68. Kessler A, Faure H, Petrel C, Rognan D, Césario M, Ruat M, et al. N1-Benzoyl-N2-[1-(-naphthyl)-trans-1, 2-diaminocyclohexanes, development of 4-chlorophenylcarboxamide (calhex 231) as a new calcium sensing receptor ligand demonstrating potent calcilytic activity. *J Med Chem* 2006;**49**:5119–28.
69. Arey BJ, Seethala R, Ma Z, Fura A, Morin J, Swartz J, et al. A novel calcium-sensing receptor antagonist transiently stimulates parathyroid hormone secretion *in vivo*. *Endocrinology* 2005;**146**:2015–22.
70. Gowen M, Stroup GB, Doods RA, James IE, Vatta BJ, Smith BR, et al. Antagonizing the parathyroid calcium receptor stimulates parathyroid hormone secretion and bone formation in osteopenic rats. *J Clin Invest* 2000;**105**:1595–604.
71. Bai M, Quinn S, Trivedi S, Kifor O, Pearce SHS, Pollak MR, et al. Expression and characterization of inactivating and activating mutations in the human Ca2+o-sensing receptor. *J Biol Chem* 1996;**271**:19537–45.
72. Pearce SHS, Bai M, Quinn SJ, Kifor O, Brown EM, Thakker RV. Functional characterization of calcium-sensing receptor mutations expressed in human embryonic kidney cells. *J Clin Invest* 1996;**98**:1860–6.
73. Bai M, Janicic N, Trivedi S, Quinn SJ, Cole DEC, Brown EM, et al. Markedly reduced activity of a mutant calcium-sensing receptor with an inserted Alu element from a kindred with familial hypocalciuric hypercalcemia and neonatal severe hyperparathyroidism. *J Clin Invest* 1997;**99**:1917–25.
74. Hauache OM. Extracellular calcium-sensing receptor: structural and functional features and association with diseases. *Braz J Med Biol Res* 2001;**34**:577–84.
75. D'Souza-Li L. The calcium-sensing receptor and related diseases. *Arq Bras Endocrinol Metabol* 2006;**50**:628–39.
76. Fan G, Goldsmith PK, Collins R, Dunn CK, Krapcho KJ, Rogers KV, et al. N-linked glycosylation of the Ca^{2+} receptor is essential for its expression at the cell surface. *Endocrinology* 1997;**138**:1916–22.
77. Ray K, Clapp P, Goldsmith PK, Spiegel AM. Identification of the sites of N-linked glycosylation on the human calcium receptor and assessment of their role in cell surface expression and signal transduction. *J Biol Chem* 1998;**273**:34558–67.
78. O'Hara PJ, Sheppard PO, Thogersen H, Venezia D, Haldeman BA, McGrane V, et al. The ligand-binding domain in metabotropic glutamate receptors is related to bacterial periplasmic binding proteins. *Neuron* 1993;**11**:41–52.
79. Kunishima N, Shimada Y, Tsuji Y, Sato T, Yamaoto M, Kumasaka T, et al. Structural basis of glutamate recognition by a dimeric metabotropic glutamate receptor. *Nature* 2000;**407**:971–7.
80. Tsuchiya D, Kunishima N, Kamiya N, Jingami H, Morikawa K. Structural views of the ligand-binding domain cores of a metabotropic glutamate receptor complexed with an antagonist and both glutamate and Gd^{3+}. *Proc Natl Acad Sci USA* 2002;**99**:2660–5.
81. Zajickova K, Vrbikova J, Canaff L, Pawelek PD, Goltzman D, Hendy GN. Identification and functional characterization of a novel mutation in the calcium-sensing receptor in familial hypocalciuric hypercalcemia: modulation of clinical severity by vitamin D status. *J Clin Endocrinol Metab* 2007;**92**:2616–23.
82. Reyes-Cruz G, Hu J, Goldsmith PK, Steinbach PJ, Spiegel AM. Human Ca^{2+} receptor extracellular domain. Analysis of function of lobe 1 loop deletion mutants. *J Biol Chem* 2001;**276**:32145–51.
83. Hauache OM, Hu J, Ray K, Spiegel AM. Functional interactions between the extracellular domain and the seven-transmembrane domain in Ca^{2+} receptor activation. *Endocrine* 2000;**13**:63–70.
84. Hu J, Hauache OM, Spiegel AM. Human Ca^{2+} receptor cysteine-rich domain. Analysis of function of mutant and chimeric receptors. *J Biol Chem* 2000;**275**:16382–9.

85. Bai M, Trivedi S, Brown EM. Dimerization of the extracellular calcium-sensing receptor (CaR) on the cell surface of CaR-transfected HEK293 cells. *J Biol Chem* 1998;**273**:23605–10.
86. Ward DT, Brown EM, Harris HW. Disulfide bonds in the extracellular calcium-polyvalent cation-sensing receptor correlate with dimer formation and its response to divalent cations in vitro. *J Biol Chem* 1998;**273**:14476–83.
87. Ray K, Hauschild BC, Steinbach PJ, Goldsmith PK, Hauache O, Spiegel AM. Identification of the cysteine residues in the amino-terminal extracellular domain of the human Ca(2+) receptor critical for dimerization. Implications for function of monomeric Ca(2+) receptor. *J Biol Chem* 1999;**274**:27642–50.
88. Zhang Z, Sun S, Quinn SJ, Brown EM, Bai M. The extracellular calcium-sensing receptor dimerizes through multiple types of intermolecular interactions. *J Biol Chem* 2001;**276**:5316–22.
89. Jiang Y, Minet E, Zhang Z, Silver PA, Bai M. Modulation of interprotomer relationships is important for activation of dimeric calcium-sensing receptor. *J Biol Chem* 2004;**279**:14147–56.
90. Fan GF, Ray K, Zhao XM, Goldsmith PK, Spiegel AM. Mutational analysis of the cysteines in the extracellular domain of the human Ca^{2+} receptor: effects on cell surface expression, dimerization and signal transduction. *FEBS Lett* 1998;**436**:353–6.
91. Jensen AA, Hansen JL, Sheikh SP, Brauner-Osborne H. Probing intermolecular protein–protein interactions in the calcium-sensing receptor homodimer using bioluminescence resonance energy transfer (BRET). *Eur J Biochem* 2002;**269**:5076–87.
92. Pidasheva S, Grant M, Canaff L, Ercan O, Kumar U, Hendy GN. The calcium-sensing receptor (CASR) dimerizes in the endoplasmic reticulum: biochemical and biophysical characterization of novel CASR mutations causing familial hypocalciuric hypercalcemia. *Hum Mol Genet* 2006;**15**:2200–9.
93. Silve C, Petrel C, Leroy C, Bruel H, Mallet E, Rognan D, et al. Delineating a Ca^{2+} binding pocket within the Venus Flytrap Module of the human calcium-sensing receptor. *J Biol Chem* 2005;**280**:37917–23.
94. Huang Y, Zhou Y, Yang W, Butters R, Lee H-W, Li S, et al. Dissection of Ca^{2+}-binding sites in the extracellular domain of the Ca^{2+}-sensing receptor. *J Biol Chem* 2007;**282**:19000–10.
95. Huang Y, Zhou Y, Castiblanco A, Yang W, Brown EM, Yang JJ. Multiple Ca^{2+}-binding sites in the extracellular domain of the Ca^{2+}-sensing receptor corresponding to cooperative Ca^{2+} response. *Biochemistry* 2009;**48**:388–98.
96. Kniazeff J, Bessis AS, Maurel D, Ansanay H, Prezeau L, Pin JP. Closed states of both binding domains of homodimeric mGlu receptors is required for full activity. *Nat Struct Mol Biol* 2004;**11**:706–13.
97. Muto T, Tsuchiya D, Morikawa K, Jingami H. Structures of the extracellular regions of the group II/II metabotropic glutamate receptors. *Proc Natl Acad Sci USA* 2007;**104**:3759–64.
98. Hu J, Reyes-Cruz G, Goldsmith PK, Gantt NM, Miller JL, Spiegel AM. Functional effects of monoclonal antibodies to the purified amino-terminal extracelular domain of the human Ca^{2+} receptor. *J Bone Miner Res* 2007;**22**:601–8.
99. Hu J, Reyes-Cruz G, Goldsmith PK, Spiegel AM. The Venus-flytrap and cysteine-rich domains of the human Ca^{2+} receptor are not linked by disulfide bonds. *J Biol Chem* 2001;**276**:6901–4.
100. Ray K, Adipetro KA, Chen C, Northup JK. Elucidation of the role of peptide linker in calcium-sensing receptor activation process. *J Biol Chem* 2007;**282**:5310–7.
101. Hu J, Reyes-Cruz G, Chen W, Jacobson KA, Spiegel AM. Identification of acidic residues in the extracellular loops of the seven-transmembrane domain of the human Ca^{2+} receptor critical for response to Ca^{2+} and a positive allosteric modulator. *J Biol Chem* 2002;**277**:46622–31.

102. Ray K, Ghosh SP, Northup JK. The role of cysteines and charged amino acids in extracellular loops of the human Ca^{2+} receptor in cell surface expression and receptor activation processes. *Endocrinology* 2004;**145**:3892–903.
103. Palczewski K, Kumasaka T, Hori T, Behnke CA, Motoshima H, Fox BA, et al. Crystal structure of rhodopsin: a G protein-coupled receptor. *Science* 2000;**289**:739–45.
104. Ray K, Tisdale J, Dodd RH, Dauban P, Ruat M, Northup JK. Calindol, a positive allosteric modulator of the human Ca^{2+}, activates an extracellular ligand-binding domain-deleted rhodopsin-like seven-transmembrane structure in the absence of Ca^{2+}. *J Biol Chem* 2005;**280**:37013–20.
105. Petrel C, Kessler A, Maslah F, Dauban P, Dodd RH, Rognan D, et al. Modeling and mutagenesis of the binding site of Calhex 231, a novel allosteric modulator of the extracellular Ca^{2+}-sensing receptor. *J Biol Chem* 2003;**278**:49487–94.
106. Petrel C, Kessler A, Dauban P, Dodd RH, Rognan D, Ruat M. Positive and negative allosteric modulators of the Ca^{2+}-sensing receptor interact within overlapping but not identical binding sites in the transmembrane domain. *J Biol Chem* 2004;**279**:18990–7.
107. Miedlich SU, Gama L, Seuwen K, Wolf RM, Breitwieser GE. Homology modeling of the transmembrane domain of the human calcium sensing receptor and localization of an allosteric binding site. *J Biol Chem* 2004;**279**:7254–63.
108. Hu J, Jiang J, Costanzi S, Thomas C, Yang W, Feyen JH, et al. A missense mutation in the seven-transmembrane domain of the human Ca^{2+} receptor conversts a negative allosteric modulator into a positive allosteric modulator. *J Biol Chem* 2006;**281**:21558–65.
109. Hu J, Spiegel AM. Naturally occurring mutations of the extracellular Ca^{2+}-sensing receptor: implications for its structure and function. *Trends Endocrinol Metab* 2003;**14**:282–8.
110. Hu J, McLarnon SJ, Mora S, Jiang J, Thomas C, Jacobson KA, et al. A region in the seven-transmembrane domain of the human Ca^{2+} receptor critical for response to Ca^{2+}. *J Biol Chem* 2005;**280**:5113–20.
111. Chang W, Chen TH, Pratt SA, Shoback D. Amino acids in the second and third intracellular loops of the parathyroid Ca^{2+}-sensing receptor mediate efficient coupling to phospholipase C. *J Biol Chem* 2000;**275**:19955–63.
112. Ray K, Fan GF, Goldsmith PK, Spiegel AM. The carboxyl terminus of the human calcium receptor: requirements for cell-surface expression and signal transduction. *J Biol Chem* 1997;**272**:31355–61.
113. Chang W, Pratt S, Chen TH, Borguignon L, Shoback D. Amino acids in the cytoplasmic C terminus of the parathyroid Ca^{2+}-sensing receptor mediate efficient cell-surface expression and phospholipase C activation. *J Biol Chem* 2001;**276**:44129–36.
114. Pi M, Oakley RH, Gesty-Palmer D, Cruickshank RD, Spurney RF, Luttrel LM, et al. β arrestin- and G protein receptor kinase-mediated calcium-sensing receptor desensitization. *Mol Endocrinol* 2005;**19**:1078–87.
115. Lorenz S, Frenzel R, Paschke R, Breitweiser GE, Miedlich SU. Functional desensitization of the extracellular calcium-sensing receptor is regulated via distinct mechanisms: role of G protein-coupled receptor kinases, protein kinase C and β-arrestins. *Endocrinology* 2007;**148**:2398–404.
116. Bai M, Trivedi S, Lane CR, Yang Y, Quinn SJ, Brown EM. Protein kinase C phosphorylation of threonine at position 888 in Ca^{2+}o-sensing receptor (CaR) inhibits coupling to Ca^{2+} store release. *J Biol Chem* 1998;**273**:21267–75.
117. Jiang Y-F, Zhang Z, Kifor O, Lane CR, Quinn SJ, Bai M. Protein kinase C (PKC) phosphorylation of the Ca^{2+}o-sensing receptor (CaR) modulates functional interaction of G proteins with the CaR cytoplasmic tail. *J Biol Chem* 2002;**277**:50543–9.
118. Bouschet T, Martin S, Henley JM. Receptor-activity-modifying proteins are required for forward trafficking of the calcium-sensing receptor to the plasma membrane. *J Cell Sci* 2005;**118**:4709–20.

119. Cetani F, Pardi E, Borsari S, Tonacchera M, Morabito E, Pinchera A, et al. Two Italian kindreds with familial hypocalciuric hypercalcemia caused by loss-of-function mutations in the calcium-sensing receptor (CaR) gene: functional characterization of a novel CaR missense mutation. *Clin Endocrinol* 2003;**58**:199–206 Erratum. Clin Endocrinol 58, 671.
120. Cetani F, Monica L, Cervia D, Borsari S, Cianferotti L, Pardi E, et al. Identification and functional characterization of loss-of-function mutations of the calcium-sensing receptor in four Italian kindreds with familial hypocalciuric hypercalcemia. *Eur J Endocrinol* 2009;**160**:481–9.
121. Huang Y, Niwa J, Sobue G, Breitweiser GE. Calcium-sensing receptor ubiquitination and degradation mediated by the E3 ubiquitin ligase dorfin. *J Biol Chem* 2006;**281**:11610–7.
122. Huang Y, Breitwieser GE. Rescue of calcium-sensing receptor mutants by allosteric modulators reveals a conformational checkpoint in receptor biogenesis. *J Biol Chem* 2007;**282**:9517–25.
123. Rus R, Haag C, Bumke-Vogt C, Bahr V, Mayr B, Mohlig M, et al. Novel inactivating mutations of the calcium-sensing receptor: the calcimimetic NPS-R-568 improves signal transduction of mutant receptors. *J Clin Endocrinol Metab* 2008;**93**:4797–803.
124. White E, McKenna J, Cavanaugh A, Breitwieser GE. Pharmacochaperone-mediated rescue of calcium-sensing receptor loss-of-function mutants. *Mol Endocrinol* 2009;**23**:1115–23.
125. Hofer AM, Brown EM. Extracellular calcium sensing and signalling. *Nat Rev Mol Cell Biol* 2003;**4**:530–8.
126. Breitweiser GE, Gama L. Calcium-sensing receptor activation induces intracellular calcium oscillations. *Am J Physiol Cell Physiol* 2001;**280**:C1412–21.
127. Young SH, Rozengurt E. Amino acids and Ca^{2+} stimulate different patterns of Ca^{2+} oscillations through the Ca^{2+}-sensing receptor. *Am J Physiol Cell Physiol* 2002;**282**:C1414–22.
128. De Luisi A, Hofer AM. Evidence that Ca^{2+} cycling by the plasma membrane Ca^{2+}-ATPase increases the 'excitability' of the extracellular Ca^{2+}-sensing receptor. *J Cell Sci* 2003;**116**:1527–38.
129. Wettschureck N, Lee E, Libutti SK, Offermanns S, Robey PG, Spiegel AM. Parathyroid-specific double knockout of Gq and G11 alpha-subunits leads to a phenotype resembling germline knockout of the extracellular Ca^{2+}-sensing receptor. *Mol Endocrinol* 2007;**21**:274–80.
130. Pi M, Chen L, Huang MZ, Luo Q, Quarles LD. Parathyroid-specific interaction of the calcium-sensing receptor and Gq. *Kidney Int* 2008;**74**:1548–56.
131. Kifor O, Diaz R, Butters R, Brown EM. The Ca^{2+}-sensing receptor (CaR) activates phospholipase C, A_2, and D in bovine parathyroid and CaR-transfected human embryonic kidney cells. *J Bone Miner Res* 1997;**12**:715–25.
132. Kifor O, Diaz R, Butters R, Kifor I, Brown EM. The calcium-sensing receptor is located in caveolin-rich plasma membrane domains of bovine parathyroid glands. *J Biol Chem* 1998;**273**:21708–13.
133. Hjalm G, MacLeod RJ, Kifor O, Chattopadhyay N, Brown EM. Filamin-A binds to the carboxyl-terminal tail of the calcium-sensing receptor, and interaction that participates in CaR-mediated activation of mitogen-activated protein kinase. *J Biol Chem* 2001;**276**:34880–7.
134. Awata H, Huang C, Handlogten ME, Miller RT. Interaction of the calcium-sensing receptor and filamin, a protein scaffolding protein. *J Biol Chem* 2001;**276**:34871–9.
135. Pi M, Spurney RF, Tu Q, Hinson T, Quarles LD. Calcium-sensing receptor activation of rho involves filamin and rho-guanine nucleotde exchange factor. *Endocrinology* 2002;**143**:3830–8.
136. Huang C, Miller RT. The calcium-sensing receptor and its interacting proteins. *J Cell Mol Med* 2007;**11**:923–34.
137. Zhang M, Breitwieser GE. High affinity interaction with filamin A protects against calcium-sensing receptor degradation. *J Biol Chem* 2005;**280**:11140–6.

138. Chang W, Chen TH, Gardner P, Shoback D. Regulation of Ca^{2+}-conducting currents in parathyroid cells by extracellular Ca^{2+} and channel blockers. *Am J Physiol* 1995;**269**:E864–77.
139. Ye C, Rogers K, Bai M, Quinn SJ, Brown EM, Vassilev PM. Agonists of the Ca^{2+}-sensing receptor activate nonselective cation channels in HEK293 cells stably transfected with the human CaR. *Biochem Biophys Res Commun* 1996;**226**:572–9.
140. Chen CJ, Barnett JV, Congo DA, Brown EM. Divalent cations suppress $3',5'$-adenosine monophosphate accumulation by stimulating a pertussis toxin-sensitive guanine nucleotide-binding protein in cultured bovine parathyroid cells. *Endocrinology* 1989;**124**:233–9.
141. Kifor O, MacLeod RJ, Diaz R, Bai M, Yamaguchi T, Yao T, et al. Regulation of MAP kinase by calcium-sensing receptor in bovine parathyroid and CaR-transfected HEK293 cells. *Am J Physiol Renal Physiol* 2001;**280**:F291–302.
142. Corbetta S, Lania A, Filopanti M, Vicentini L, Ballaré E, Spada A. Mitogen-activated protein kinase cascade in human normal and tumoral parathyroid cells. *J Clin Endocrinol Metab* 2002;**87**:2201–5.
143. Bourdeau A, Souberbielle JC, Bonnet P, Herviaux P, Sachs C, Lieberherr M. Phospholipase-A_2 action and arachidonic acid metabolism in calcium-mediated parathyroid hormone secretion. *Endocrinology* 1992;**130**:1339–44.
144. Bourdeau A, Moutahir M, Souberbielle JC, Bonnet P, Herviaux P, Sachs C, et al. Effects of lipoxygenase products of arachidonate metabolism on parathyroid hormone secretion. *Endocrinology* 1994;**135**:1109–12.
145. Quinn SJ, Kifor O, Kifor I, Butters Jr RR, Brown EM. Role of the cytoskeleton in extracellular calcium-regulated PTH release. *Biochem Biophys Res Commun* 2007;**354**:8–13.
146. Kifor O, Kifor I, Moore Jr FD, Butters Jr RR, Brown EM. m-Calpain colocalizes with the calcium-sensing receptor (CaR) in caveolae in parathyroid cells and participates in degradation of the CaR. *J Biol Chem* 2003;**278**:31167–76.
147. Brookman JJ, Farrow SM, Nicholson L, O'Riordan JLH, Hendy GN. Regulation by calcium of parathyroid hormone mRNA in cultured parathyroid tissue. *J Bone Miner Res* 1987;**6**:529–37.
148. Levi R, Ben-Dov IZ, Lavi-Moshayoff V, Dinur M, Martin D, Naveh-Many T, et al. Increased parathyroid hormone gene expression in secondary hyperparathyroidism of experimental uremia is reversed by calcimetics: correlation with posttranslational modification of the trans acting factor AUF1. *J Am Soc Nephrol* 2006;**17**:107–12.
149. Ho C, Conner DA, Pollak MR, Ladd DJ, Kifor O, Warren HB, et al. A mouse model of human familial hypocalciuric hypercalcemia and neonatal severe hyperparathyroidism. *Nat Genet* 1995;**11**:389–94.
150. Wada M, Ishii H, Furuya Y, Fox J, Nemeth EF, Nagano N. NPS R-568 halts or reverses osteitis fibrosa in uremic rats. *Kidney Int* 1998;**53**:448–53.
151. Colloton M, Shatzen E, Miller G, StehmanpBreen C, Wada M, Lacey D, et al. Cinacalcet Hcl attenuates parathyroid hyperplasia in a rat model of secondary hyperparathyroidism. *Kidney Int* 2005;**67**:467–76.
152. Panda DK, Miao D, Bolivar I, Li J, Juo R, Hendy GN, et al. Inactivation of the 25-hydroxyvitamin D-1-hydroxylase and vitamin D receptor demonstrates independent and interdependent effects of calcium and vitamin D on skeletal and mineral homeostasis. *J Biol Chem* 2004;**279**:16754–66.
153. Friedman PA, Gesek FA. Cellular calcium transport in renal epithelia: measurement, mechanisms, and regulation. *Physiol Rev* 1995;**75**:429–71.
154. Riccardi D, Hall AE, Chattopadhyay N, Xu JZ, Brown EM, Hebert SC. Localization of the extracellular Ca^{2+}/polyvalent cation-sensing protein in rat kidney. *Am J Physiol* 1998;**274**:F611–22.
155. Weisinger JR, Favus MJ, Langman CB, Bushinsky DA. Regulation of 1, 25-dihydroxyvitamin D3 by calcium in the parathyroidectomized, parathyroid hormone-replete rat. *J Bone Miner Res* 1989;**4**:929–35.

156. Quamme GA, de Rouffignac C. Epithelial magnesium transport and regulation by the kidney. *Front Biosci* 2000;**5**:D694–711.
157. Wang WH, Lu M, Hebert SC. Cytochrome P-450 metabolites mediate extracellular Ca^{2+}-induced inhibition of apical K^+ channels in the TAL. *Am J Physiol* 1996;**271**:C103–11.
158. Ba J, Friedman PA. Calcium-sensing receptor regulation of renal mineral ion transport. *Cell Calcium* 2004;**35**:229–37.
159. Thebault S, Hoenderop JG, Bindels RJ. Epithelial Ca^{2+} and Mg^{2+} channels in kidney disease. *Adv Chronic Kidney Dis* 2006;**13**:110–7.
160. Huang C, Miller RT. Regulation of renal ion transport by the calcium-sensing receptor: an update. *Curr Opin Nephrol Hypertens* 2007;**16**:437–43.
161. Clemens TL, McGlade SA, Garrett KP, Craviso GL, Hendy GN. Extracellular calcium modulates vitamin D-dependent calbindin-D_{28k} gene expression in chick kidney cells. *Endocrinology* 1989;**124**:1582–4.
162. Bapty BW, Dai LJ, Ritchie G, Jirik F, Canaff L, Hendy GN, et al. Extracellular Mg^{2+} and Ca^{2+}-sensing in mouse distal convoluted tubule cells. *Kidney Int* 1998;**53**:583–92.
163. Bapty BW, Ritchie G, Canaff L, Hendy GN, Quamme GA. Extracellular Mg^{2+} and Ca^{2+}-sensing inhibits hormone-stimulated Mg^{2+} uptake in mouse distal convoluted tubule cells. *Am J Physiol* 1998;**275**:F353–60.
164. Jackson CE, Boonstra C. Hereditary hypercalcemia and parathyroid hyperplasia without definite hyperparathyroidism. *J Lab Clin Med* 1966;**68**:883 Abs 862.
165. Foley Jr TP, Harrison HC, Arnaud CD, Harrison HE. Familial benign hypercalcemia. *J Pediatr* 1972;**81**:1060–7.
166. Law Jr WM, Heath III H. Familial benign hypercalcemia (hypocalciuric hypercalcemia): clinical and pathogenetic stucy of 21 families. *Ann Intern Med* 1985;**102**:511–9.
167. Marx SJ, Attie MF, Levine MA, Spiegel AM, Downs RWJ, Lasker RD. The hypocalciuric or benign variant of familial hypercalcemia: clinical and biochemical features of fifteen families. *Medicine (Baltimore)* 1981;**60**:397–412.
168. Heath III H. Familial benign hypercalcemia, a troublesome mimic of mild primary hyperparathyroidism. *Endocrinol Metab Clin N Am* 1989;**18**:723–40.
169. Kristiansen JH, Rodbro P, Christiansen C, Johansen J, Jensen JT. Familial hypocalciuric hypercalcemia III: Bone mineral metabolism. *Clin Endocrinl (Oxf)* 1987;**26**:713–6.
170. Christensen SE, Nissen PH, Vestergaard P, Heickendorff L, Rejnmark L, Brixen K, et al. Skeletal consequences of familiar hypocalciuric hypercalcemia versus primary hyperparathyroidism. *Clin Endocrinol (Oxf)* 2009; Epub ahead of print.
171. Davies M, Klimiuk PS, Adams PH, Lumb GA, Large DM, Anderson DC. Familial hypocalciuric hypercalcemia and acute pancraetitis. *Br Med J* 1981;**282**:1023–5.
172. Wang D, Canaff L, Davidson D, Corluka A, Liu H, Hendy GN, et al. Alterations in the sensing and transport of phosphate and calcium by differentiating chondrocytes. *J Biol Chem* 2001;**276**:33995–4005.
173. Marx SJ, Spiegel AM, Brown EM, Koehler JO, Gardner DG, Brennan MF, et al. Divalent cation metabolism. Familial hypocalciuric hypercalcemia versus typical primary hyperparathyroidism. *Am J Med* 1978;**65**:235–42.
174. Auwerx J, Demedts M, Bouillon R. Altered parathyroid set point to calcium in familial hypocalciuric hypercalcemia. *Acta Endocrinol* 1984;**106**:215–8.
175. Khosla S, Ebeling PR, Firek AF, Burritt MM, Kao PC, Heath III H. Calcium infusion suggests a "set-point" abnormality of parathyroid gland function in familial benign hypercalcemia and more complex disturbances in primary hyperparathyroidism. *J Clin Endocrinol Metab* 1993;**76**:715–20.
176. Gunn I, Wallace J. Urine calcium and serum ionized calcium, total calcium and parathyroid hormone concentrations in the diagnosis of primary hyperparathyroidism and familial benign hypercalcemia. *Ann Clin Biochem* 1992;**29**:52–8.

177. Paterson CR, Gunn A. Familial benign hypercalcemia. *Lancet* 1981;**1**:61–3.
178. Law Jr WM, Carney JA, Heath III H. Parathyroid glands in familial benign hypercalcemia (familial hypocalciuric hypercalcemia). *Am J Med* 1984;**76**:1021–6.
179. Thogeirsson U, Costa J, Marx SJ. The parathyroid glands in familial benign hypocalciuric hypercalcemia. *Hum Pathol* 1981;**12**:229–37.
180. Attie MF, Gill Jr J, Stock JL, Spiegel AM, Downs Jr RW, Levine M, et al. Urinary calcium excretion in familial hypocalciuric hypercalcemia. Persistence of relative hypocalciuria after induction of hypoparathyroidism. *J Clin Invest* 1983;**72**:667–76.
181. Davies M, Adams PH, Lumb GA, Berry J, Loveridge N. Familial hypocalciuric hypercalcemia: evidence for continued enhanced renal tubular reabsorption of calcium following total parathyroidectomy. *Acta Endocrinol (Copenhagen)* 1984;**106**:499–504.
182. Menko FH, Bijvoet OLM, Fronen JLHH, Sandler LM, Adami S, O'Riordan JLH, et al. Familial benign hypercalcemia: study of a large family. *Q J Med* 1983;**206**:120–40.
183. Pasieka J, Anderson M, Hanley D. Familial benign hypercalcemia: hypercalciuria and hypocalciuria in affected members of a small kindred. *Clin Endocrinol* 1990;**33**:429–33.
184. Pearce SHS, Wooding C, Davies M, Tollefsen SE, Whyte MP, Thakker RV. Calcium-sensing receptor mutations in familial hypocalciuric hypercalcaemia with recurrent pancreatitis. *Clin Endocrinol* 1996;**45**:675–80.
185. Carling T, Szabo E, Bai M, Ridefelt P, Westin G, Gustavsson P, et al. Familial hypercalcemia and hypercalciuria caused by a novel mutation in the cytoplasmic tail of the calcium receptor. *J Clin Endocrinol Metab* 2000;**85**:2042–7.
186. Marx SJ, Attie MF, Stock JL, Spiegel AM, Levine MA. Maximal urine-concentrating ability: familial hypocalciuric hypercalcemia versus typical primary hyperparathyroidism. *J Clin Endocrinol Metab* 1981;**52**:736–40.
187. Sands JM, Naruse M, Baum M, Jo I, Hebert SC, Brown EM, et al. Apical extracellular calcium/polyvalent cation sensing receptor regulates vasopressin-elicited water permeability in rat kidney inner medullary collecting duct. *J Clin Invest* 1997;**99**:1399–405.
188. Pratt E, Geren B, Neuhauser E. Hypercalcemia and idiopathic hyperplasia of the parathyroid glands in an infant. *J Pediatr (St Louis)* 1947;**30**:388–99.
189. Hillman DA, Scriver CR, Pedvis S, Shragovitch I. Neonatal familial primary hyperparathyroidism. *N Engl J Med* 1964;**270**:810–1.
190. Marx SJ, Attie MF, Spiegel AM, Levine MA, Lasker RD, Fox M. An association between neonatal severe primary hyperparathyroidism and familial hypocalciuric hypercalcemia in three kindreds. *N Engl J Med* 1982;**306**:257–64.
191. Marx SJ, Fraser D, Rapoport A. Familial hypocalciuric hypercalcemia: mild expression of the gene in heterozygotes and severe expression in homozygotes. *Am J Med* 1985;**78**:15–22.
192. Marx SJ, Lasker RD, Brown EM, Fitzpatrick LA, Sweeney NB, Goldbloom RB, et al. Secretory dysfunction in parathyroid cells from a neonate with severe primary hyperparathyroidism. *J Clin Endocrinol Metab* 1986;**62**:445–9.
193. Cole DEC, Forsythe CR, Dooley JM, Grantmyre EB, Salisbury SR. Primary neonatal hyperparathyroidism: a devastating neurodevelopmental disorder if left untreated. *J Craniofac Genet Dev Biol* 1990;**10**:205–14.
194. Cole DEC, Janicic N, Salisbury SR, Hendy GN. Neonatal severe hyperparathyroidism, secondary hyperparathyroidism, and familial hypocalciuric hypercalcemia: multiple different phenotypes associated with an inactivating Alu insertion mutation of the calcium-sensing receptor (CASR) gene. *Am J Med Genet* 1997;**71**:202–10.
195. Pearce S, Steinmann B. Casting new light on the clinical spectrum of neonatal severe hyperparathyroidism. *Clin Endocrinol* 1999;**50**:691–3.
196. Janicic N, Pausova Z, Cole DEC, Hendy GN. Insertion of an Alu sequence in the Ca^{2+}-sensing receptor gene in familial hypocalciuric hypercalcemia and neonatal severe hyperparathyroidism. *Am J Hum Genet* 1995;**56**:880–6.

197. Kobayashi M, Tanaka H, Tsuzuki K, Tsuzuki M, Igaki H, Ichinose Y, et al. Two novel missense mutations in calcium-sensing receptor gene associated with neonatal severe hyperparathyroidism. *J Clin Endocrinol Metab* 1997;**82**:2716–9.
198. Bai M, Pearce SHS, Kifor O, Trivedi S, Stauffer UG, Thakker RV, et al. In vivo and in vitro characterization of neonatal hyperparathyroidism resulting from a *de novo*, heterozygous mutation in the Ca^{2+}-sensing receptor gene: normal maternal calcium homeostasis as a cause of secondary hyperparathyroidism in familial benign hypocalciuric hypercalcemia. *J Clin Invest* 1997;**99**:88–96.
199. Kovacs CS. Calcium and bone metabolism in pregnancy and lactation. *J Clin Endocrinol Metab* 2001;**86**:2344–8.
200. Pagan YL, Hirschhorn J, Yang B, D'Souza-Li L, Majzoub JA, Hendy GN. Maternal activating mutation of the calcium-sensing receptor: implications for calcium metabolism in the neonate. *J Pediatr Endocrinol Metab* 2004;**17**:673–7.
201. Chou YH, Brown EM, Levi T, Crowe G, Atkinson AB, Arnqvist H, et al. The gene responsible for familial hypocalciuric hypercalcemia maps to chromosome 3q in four unrelated families. *Nat Genet* 1992;**1**:295–300.
202. Heath III H, Jackson CE, Otterud B, Leppert MF. Genetic linkage analysis of familial benign (hypocalciuric) hypercalcemia: evidence for locus heterogeneity. *Am J Hum Genet* 1993;**53**:193–200.
203. Pearce SHS, Trump D, Wooding C, Besser GM, Chew SL, Grant DB, et al. Calcium-sensing receptor mutations in familial benign hypercalcemia and neonatal hyperparathyroidism. *J Clin Invest* 1995;**96**:2683–92.
204. McMurtry CT, Schranck FW, Walkenhorst DA, Murphy WA, Kocher DB, Teitelbaum SL, et al. Significant developmental elevation in serum parathyroid hormone levels in a large kindred with familial benign (hypocalciuric) hypercalcemia. *Am J Med* 1992;**93**:247–58.
205. Trump D, Whyte MP, Wooding C, Pang JT, Pearce SHS, Kocher DB, et al. Linkage studies in a kindred from Oklahoma, with familial benign (hypocalciuric) hypercalcemia (FBH) and developmental elevations in serum parathyroid hormone levels, indicate a third locus for FBH. *Hum Genet* 1995;**96**:183–7.
206. Lloyd SE, Pannett AAJ, Dixon PH, Whyte MP, Thakker RV. Localization of familial benign hypercalcemia, Oklahoma variant (FBH_{ok}), to chromosome 19q13. *Am J Hum Genet* 1999;**64**:189–95.
207. Hendy GN, Cole DEC. Parathyroid disorders. In: Rimoin DL, Connor JM, Pyeritz RE, Korf B, editors. *Principles and practice of medical genetics*. 5th ed London: Churchill Livingstone; 2007. p. 1951–79.
208. Datta R, Waheed A, Shah GN, Sly WS. Signal sequence mutation in autosomal dominant form of hypoparathyroidism induces apoptosis that is corrected by a chemical chaperone. *Proc Natl Acad Sci USA* 2007;**104**:19989–94.
209. Mannstadt M, Bertrand G, Muresan M, Weryha G, Leheup B, Pulusani SR, et al. Dominant-negative GCMB mutations cause an autosomal dominant form of hypoparathyroidism. *J Clin Endocrinol Metab* 2008;**93**:3568–76.
210. Heath DA. Familial hypocalcemia—not hypoparathyroidism. *N Engl J Med* 1996;**335**:1144–5.
211. Vargas-Poussou R, Huang C, Hulin P, Houillier P, Jeunemaitre X, Paillard M, et al. Functional characterization of a calcium-sensing receptor mutation in severe autosomal dominant hypocalcemia with a Bartter-like syndrome. *J Am Soc Nephrol* 2002;**13**:2259–66.
212. Watanabe S, Fukumoto S, Chang H, Takeuchi Y, Hasegawa Y, Okazaki R, et al. Association between activating mutations of calcium-sensing receptor and Bartter's syndrome. *Lancet* 2002;**360**:692–4.
213. Vezzoli G, Arcidiacono T, Paloschi V, Terranegra A, Biasion R, Weber G, et al. Autosomal dominant hypocalcemia with mild type 5 Bartter syndrome. *J Nephrol* 2006;**19**:525–8.

214. Hebert SC. Bartter syndrome. *Curr Opin Nephrol Hypertens* 2003;**12**:527–32.
215. Hendy GN, D'Souza-Li L, Yang B, Canaff L, Cole DEC. Mutations in the calcium-sensing receptor (CASR) in familial hypocalciuric hypercalcemia, neonatal severe hyperparathyroidism, and autosomal dominant hypocalcemia. *Hum Mutat* 2000;**16**:281–96.
216. Pidasheva S, D'Souza-Li L, Canaff L, Cole DEC, Hendy GN. *CASR*db: calcium-sensing receptor locus-specific database for mutations causing familial (benign) hypocalciuric hypercalcemia, neonatal severe hyperparathyroidism, and autosomal dominant hypocalcemia. *Hum Mutat* 2004;**24**:107–11.
217. Heath III H, Odelberg S, Jackson CE, Teh BT, Hayward N, Larsson C, et al. Clustered inactivation mutations and benign polymorphisms of the calcium receptor gene in familial benign hypocalciuric hypercalcemia suggest receptor functional domains. *J Clin Endocrinol Metab* 1996;**81**:1312–7.
218. D'Souza-Li L, Yang B, Canaff L, Bai M, Hanley DA, Bastepe M, et al. Identification and functional characterization of novel calcium-sensing receptor mutations in familial hypocalciuric hypercalcemia and autosomal dominant hypocalcemia. *J Clin Endocrinol Metab* 2002;**87**:1309–18.
219. Bai M, Trivedi S, Kifor O, Quinn SJ, Brown EM. Intermolecular interaction between dimeric calcium-sensing receptor monomers are important for its normal function. *Proc Natl Acad Sci USA* 1999;**96**:2834–9.
220. Wystrychowski A, Pidasheva S, Canaff L, Chudek J, Kokot F, Wiecek A, et al. Functional characterization of calcium-sensing receptor codon 227 mutations presenting either as familial (benign) hypocalciuric hypercalcemia or neonatal hyperparathyroidism. *J Clin Endocrinol Metab* 2005;**90**:864–70.
221. De Andrade SC, Kohara SK, D'Souza-Li L. Novel mutation of the calcium-sensing receptor gene in familial hypocalciuric hypercalcemia and neonatal severe hyperparathyroidism. *Clin Endocrinol* 2006;**65**:826–31.
222. Miyashiro K, Kunii I, Manna TD, De Menezes Filho HC, Damiani D, Setian N, et al. Severe hypercalcemia in a 9-year-old Brazilian girl due to a novel inactivating mutation of the calcium-sensing receptor. *J Clin Endocrinol Metab* 2004;**89**:5936–41.
223. Zhang Z, Qiu W, Quinn SJ, Conigrave AD, Brown EM, Bai M. Three adjacent serines in the extracellular domain of the CaR are required for L-amino acid-mediated potentiation of receptor function. *J Biol Chem* 2002;**277**:33727–35.
224. Hoff AO, Cote GJ, Fritsche Jr HA, Qiu H, Schultz PN, Gagel RF. Calcium-induced activation of a mutant G-protein-coupled receptor causes *in vitro* transformation of NIH/3T3 cells. *Neoplasia* 1999;**1**:485–91.
225. Cole DEC, Yun FHJ, Wong BYL, Shuen AY, Booth RA, Scillitani A, et al. Calcium-sensing receptor (*CASR*) mutations and denaturing high performance liquid chromatography (DHPLC). *J Mol Endocrinol* 2009;**42**:331–9.
226. Kifor O, Moore Jr FD, Delaney M, Garber J, Hendy GN, Butters R, et al. A syndrome of hypocalciuric hypercalcemia caused by autoantibodies directed at the calcium-sensing receptor. *J Clin Endocrinol Metab* 2003;**88**:60–72.
227. Pallais JC, Kifor O, Cheri YB, Slovik D, Brown EM. Acquired hypocalciuric hypercalcemia due to autoantibodies against the calcium-sensing receptor. *N Engl J Med* 2004;**351**:362–9.
228. Makita N, Sato J, Manaka K, Shoji Y, Oishi A, Hashimoto M, et al. An acquired hypocalciuric hypercalcemia autoantibody induces allosteric transition among active human Ca-sensing receptor conformations. *Proc Natl Acad Sci USA* 2007;**104**:5443–8.
229. Brown EM. Anti-parathyroid and anti-calcium-sensing receptor antibodies in autoimmune hypoparathyroidism. *Endocrinol Metab Clin N Am* 2009;**38**:437–45.
230. Posillico JT, Wortsman J, Srikanta S, Eisenbarth GS, Mallette LE, Brown EM. Parathyroid cell surface autoantibodies that inhibit parathyroid hormone secretion from dispersed human parathyroid cells. *J Bone Miner Res* 1986;**1**:475–83.

231. Li Y, Song YH, Rais N, Connor E, Schatz D, Muir A, et al. Autoantibodies to the extracellular domain of the calcium-sensing receptor in patients with acquired hypoparathyroidism. *J Clin Invest* 1996;**97**:910–4.
232. Kifor O, McElduff A, Leboff MS, Moore Jr FD, Butters R, Gao P, et al. Activating antibodies to the calcium-sensing receptor in two patients with autoimmune hypoparathroidism. *J Clin Endocrinol Metab* 2004;**89**:548–56.
233. Goswami R, Brown EM, Kochupillai N, Gupta N, Rani R, Kifor O, et al. Prevalence of calcium sensing receptor autoantibodies in patients with sporadic idiopathic hypoparathyroidism. *Eur J Endocrinol* 2004;**150**:9–18.
234. Mayer A, Ploix C, Orgiazzi J, Desbos A, Moreira A, Vidal H, et al. Calcium-sensing receptor autoantibodies are relevant markers of acquired hypoparathyroidism. *J Clin Endocrinol Metab* 2004;**89**:4484–8.
235. Gavalas NG, Kemp EH, Krohn KJE, Brown EM, Watson PF, Weetman AP. The calcium-sensing receptor is a target of autoantibodies in patients with autoimmune polyendocrine syndrome type 1. *J Clin Endocrinol Metab* 2007;**92**:2107–14.
236. Scillitani A, Guarnieri V, De Geronimo S, Muscarella LA, Battista C, D'Agruma L, et al. Blood ionized calcium is associated with clustered polymorphisms in the carboxyl-terminal tail of the calcium-sensing receptor. *J Clin Endocrinol Metab* 2004;**89**:5634–8.
237. Scillitani A, Guarnieri V, Battista C, De Geronimo S, Muscarella LA, Chiodini I, et al. Primary hyperparathyroidism and the presence of kidney stones are associated with different haplotypes of the calcium-sensing receptor. *J Clin Endocrinol Metab* 2007;**92**:277–83.
238. Cole DEC, Peltekova VD, Rubin LA, Hawker GA, Vieth R, Liew CC, et al. A986S polymorphism of the calcium-sensing receptor and circulating calcium concentrations. *Lancet* 1999;**353**:112–5.
239. Cole DE, Vieth R, Trang HM, Wong BY, Hendy GN, Rubin LA. Association between total serum calcium and the A986S polymorphism of the calcium-sensing receptor gene. *Mol Genet Metab* 2001;**72**:168–74.
240. Lorentzon M, Lorentzon R, Lerner UH, Nordström P. Calcium sensing receptor gene polymorphism, circulating calcium concentrations and bone mineral density in healthy adolescent girls. *Eur J Endocrinol* 2001;**144**:257–61.
241. Marz W, Seelhorst U, Wellnitz B, Tiran B, Obermayer-Pitsch B, Renner W, et al. Alanine to serine polymorphism at position 986 of the calcium-sensing receptor associated with coronary heart disease, myocardial infarction, all-cause, and cardiovascular mortality. *J Clin Endocrinol Metab* 2007;**92**:2363–9.
242. Kelly C, Gunn IR, Gaffney D, Devgun MS. Serum calcium, urine calcium and polymorphisms of the calcium-sensing receptor gene. *Ann Clin Biochem* 2006;**43**:503–6.
243. Miedlich S, Lamesch P, Mueller A, Paschke R. Frequency of the calcium-sensing receptor variant A986S in patients with primary hyperparathyroidism. *Eur J Endocrinol* 2001;**145**:421–7.
244. Yamauchi M, Sugimoto T, Yamaguchi T, Yano S, Kanzawa M, Kobayashi A, et al. Association of polymorphic alleles of the calcium-sensing receptor gene with the clinical severity of primary hyperparathyroidism. *Clin Endocrinol (Oxf)* 2001;**55**:373–9.
245. Cetani F, Borsari S, Vignali E, Pardi E, Picone A, Cianferotti L, et al. Calcium-sensing receptor gene polymorphisms in primary hyperparathyroidism. *J Endocrinol Invest* 2002;**25**:614–9.
246. Bollerslev J, Wilson SG, Dick IM, Devine A, Dhaliwal SS, Prince RL. Calcium-sensing receptor gene polymorphism A986S does not predict serum calcium level, bone mineral density, calcaneal ultrasound indices, or fracture rate in a large cohort of elderly women. *Calcif Tissue Int* 2004;**74**:12–7.
247. Petrucci M, Scott P, Ouimet D, Trouvé ML, Proulx Y, Valiquette L, et al. Evaluation of the calcium-sensing receptor gene in idiopathic hypercalciuria and calcium nephrolithiasis. *Kidney Int* 2000;**58**:38–42.

248. Young R, Wu F, Van de Water N, Ames R, Gamble G, Reid IR. Calcium sensing receptor gene A986S polymorphism and responsiveness to calcium supplementation in postmenopausal women. *J Clin Endocrinol Metab* 2003;**88**:697–700.
249. Harding B, Curley AJ, Hannan FM, Christie PT, Bowl MR, Turner JJ, et al. Functional characterization of calcium-sensing receptor polymorphisms and absence of association with indices of calcium and bone mineral density. *Clin Endocrinol* 2006;**65**:598–605.
250. Tsukamoto K, Orimo H, Hosoi T, Miyao M, Ota N, Nakajima T, et al. Association of bone mineral density with polymorphism of the human calcium-sensing receptor locus. *Calcif Tissue Int* 2000;**66**:181–3.
251. Lips MA, Syddall HE, Gaunt TR, Rodriguez S, Day IN, Cooper C, et al. Interaction between birthweight and polymorphism in the calcium-sensing receptor gene indetermination of adult bone mass: the Hertfordshire cohort study. *J Rheumatol* 2007;**34**:769–75.
252. Katsumata K, Nishizawa K, Unno A, Fujita Y, Tokita A. Association of gene polymorphisms and bone density in Japanese girls. *J Bone Miner Res* 2002;**20**:164–9.
253. Takacs I, Speer G, Bajnok E, Tabak A, Nagy Z, Horvath C, et al. Lack of association between calcium-sensing receptor gene A986S polymorphism and bone mineral density in Hungarian postmenopausal women. *Bone* 2002;**30**:849–52.
254. Cetani F, Pardi E, Borsari S, Vignali E, Dipollina G, Braga V, et al. Calcium-sensing receptor gene polymorphism is not associated with bone mineral density in Italian postmenopausal women. *Eur J Endocrinol* 2003;**148**:603–7.
255. Lau HH, Ng MY, Cheung WM, Paterson AD, Sham PC, Luk KD, et al. Assessment of linkage and association of 13 genetic loci with bone mineral density. *J Bone Miner Metab* 2006;**24**:226–34.
256. Cole DEC, Trang HM, Vieth R, Peltekova VD, Pierratos A, Wong BYL, et al. Calcium excretion is independently associated with common polymorphisms of the vitamin D receptor (VDR) and calcium-sensing receptor (CASR) genes in a nephrolithiasis population. *Bone* 1998;**23S**:T195.
257. Corbetta S, Eller-Vainicher C, Filopanti M, Saeli P, Vezzoli G, Arcidiacono T, et al. R990G polymorphism of the calcium-sensing receptor and renal calcium excretion in patients with primary hyperparathyroidism. *Eur J Endocrinol* 2006;**155**:687–92.
258. Vezzoli G, Tanini A, Ferrucci L, Soldati L, Bianchin C, Franceschelli F, et al. Influence of calcium-sensing receptor gene on urinary calcium excretion in stone-forming patients. *J Am Soc Nephrol* 2002;**13**:2517–23.
259. Vezzoli G, Terranegra A, Arcidiacono T, Biasion R, Coviello D, Syren ML, et al. R990G polymorphism of calcium-sening receptor does produce a gain-of-function and predispose to primary hypercalciuria. *Kidney Int* 2007;**71**:1155–62.
260. Tobin MD, Tomaszewski M, Braund PS, Hajat C, Raleigh SM, Palmer TM, et al. Common variants in genes underlying monogenic hypertension and hypotension and blood pressure in the general population. *Hypertension* 2008;**51**:1658–64.
261. Jung J, Foroud TM, Eckert GJ, Flury-Wetherill L, Edenberg HL, Xuei X, et al. Association of the calcium-sensing receptor gene with blood pressure and urinary calcium in African-Americans. *J Clin Endocrinol Metab* 2009;**94**:1042–8.
262. Speer G, Cseh K, Mucsi K, Takacs I, Dworak O, Winkler G, et al. Calcium-sensing receptor A986S polymorphism in human rectal cancer. *Int J Colorectal Dis* 2002;**17**:20–4.
263. Peters U, Chatterjee N, Yeager M, Chanock SJ, Schoen RE, McGlynn KA, et al. Association of genetic variants in the calcium-sensing receptor with risk of colorectal adenoma. *Cancer Epidemiol Biomarkers Prev* 2004;**13**:2181–6.
264. Dong LM, Ulrich CM, Hsu L, Duggan DJ, Benitez DS, White E, et al. Genetic variation in calcium-sensing receptor and risk for colon cancer. *Cancer Epidemiol Biomarkers Prev* 2008;**17**:2755–65.

265. Muddana V, Lamb J, Greer JB, Elinoff B, Hawes RH, Cotton PB, et al. Association between calcium-sensing receptor gene polymorphisms and chronic pancreatitis in a US population: role of serine protease inhibitor Kazal1 type and alcohol. *World J Gastroenterol* 2008;**14**:4486–91.
266. Conley YP, Mukherjee A, Kammerer C, Dekosky ST, Kamboh MI, Finegold DN, et al. Evidence supporting a role for the calcium-sensing receptor in Alzheimer disease. *Am J Med Genet B Neuropsychiatr Genet* 2009;**150B**:703–9.
267. Soie YL, Karperien M, Bakker B, Breuning MH, Hendy GN, Papapoulos SE. Familial benign hypercalcemia (FBH) with age-associated hypercalciuria and a missense mutation in the calcium-sensing receptor (CaSR) expands the spectrum of the syndrome towards primary hyperparathyroidism. *J Bone Miner Res* 1999;**14s1**:S447, SU062.
268. Simonds WF, James-Newton LA, Agarwal SK, Yang B, Skarulis MC, Hendy GN, et al. Familial isolated hyperparathyroidism. Clinical and genetic characteristics of 36 kindreds. *Medicine* 2002;**81**:1–26.
269. Warner J, Epstein M, Sweet A, Singh D, Burgess J, Stranks S, et al. Genetic testing in familial isolated hyperparathyroidism: unexpected results and their implications. *J Med Genet* 2004;**41**:155–60.
270. Hosokawa Y, Pollak MR, Brown EM, Arnold A. Mutational analysis of the extracellular Ca^{2+}-sensing receptor gene in human parathyroid tumors. *J Clin Endocrinol Metab* 1995;**80**:3107–10.
271. Thompson DB, Samowitz WS, Odelberg S, Davis RK, Szabo J, Heath III H. Genetic abnormalities in sporadic parathyroid adenomas: loss of heterozygosity for chromosome 3q markers flanking the calcium receptor locus. *J Clin Endocrinol Metab* 1995;**80**:3377–80.
272. Cetani F, Pinchera A, Pardi E, Cianferotti L, Vignali E, Picone A, et al. No evidence for mutations in the calcium-sensing receptor gene in sporadic parathyroid adenomas. *J Bone Miner Res* 1999;**14**:878–82.
273. Hendy GN, Arnold A. Molecular basis of PTH overexpression. In: Bilezikian JP, Martin TJ, Raisz L, editors. *Principles of Bone Biology*. 3rd ed San Diego: Academic Press; 2008. p. 1311–26.
274. Gogusev J, Duchambon P, Hory B, Giovannini M, Goureau Y, Sarfait E, et al. Depressed expression of calcium receptor in parathyroid gland tissue of patients with hyperparathyroidism. *Kidney Int* 1997;**51**:328–36.
275. Kremer R, Bolivar I, Goltzman D, Hendy GN. Influence of calcium and 1, 25-dihydroxycholecalciferol on proliferation and proto-oncogene expression in primary cultures of bovine parathyroid cells. *Endocrinology* 1989;**125**:935–41.
276. Yao JJ, Bai S, Karnauskas AJ, Bushinsky DA, Favus MJ. Regulation of renal calcium receptor gene expression by 1, 25-dihydroxyvitamin D3 in genetic hypercalciuric stone-forming rats. *J Am Soc Nephrol* 2005;**16**:1300–8.
277. Bai S, Favus M. Vitamin D and calcium receptors: links to hypercalciuria. *Curr Opin Nephrol Hypertens* 2006;**15**:381–5.
278. Chakravarti B, Dwivedi SKD, Mithal A, Chattopadhyay N. Calcium-sensing receptor in cancer: good cop or bad cop? *Endocrine* 2008;**35**:271–84.
279. Saidak Z, Mentaverri R, Brown EM. The role of the calcium-sensing receptor in the development and progression of cancer. *Endocrine Rev* 2009;**30**:178–95.
280. Chakrabarty S, Wang H, Canaff L, Hendy GN, Appelman H, Varani J. Calcium-sensing receptor in human colon carcinoma: interaction with Ca^{2+} and 1, 25-dihydroxyvitamin D. *Cancer Res* 2005;**65**:493–8.
281. Sanders JL, Chattopadhyay N, Kifor O, Yamaguchi T, Butters RR, Brown EM. Extracellular calcium-sensing receptor expression and its potential role in regulating parathyroid hormone-related peptide secretion in human breast cancer cell lines. *Endocrinology* 2000;**141**:4357–64.

282. Sanders JL, Chattopadhyay N, Kifor O, Yamaguchi T, Brown EM. Ca(2+)-sensing receptor expression and PTHrP secretion in PC-3 human prostate cancer cells. *Am J Physiol Endocrinol Metab* 2001;**281**:E1267–74.
283. Yano S, Macleod RJ, Chattopadhyay N, Tfelt-Hansen J, Kifor O, Butters RR, et al. Calcium-sensing receptor activation stimulates parathyroid hormone-related protein secretion in prostate cancer cells: role of epidermal growth factor receptor transactivation. *Bone* 2004;**35**:664–72.
284. Zaloga GP. Hypocalcema in critically ill patients. *Crit Care Med* 1992;**20**:251–62.
285. Zivin JR, Gooley T, Zager RA, Ryan MJ. Hypocalcemia: a pervasive metabolic abnormality in the critically ill. *Am J Kidney Dis* 2001;**37**:689–98.
286. Drüeke TB, Ritz E. Treatment of secondary hyperparathyroidism in CKD patients with cinacalcet and/or vitamin D derivatives. *Clin J Am Soc Nephrol* 2009;**4**:234–41.
287. Wetmore JB, Quarles LD. Calcimimetics or vitamin D analogs for suppressing parathyroid hormone in end-stage renal disease: time for a paradigm shift? *Nat Clin Pract Nephrol* 2009;**5**:24–33.
288. Khan A, Grey A, Shoback D. Medical management of asymptomatic primary hyperparathyroidism: proceedings of the third international workshop. *J Clin Endocrinol Metab* 2009;**94**:373–81.
289. Nemeth EF. Drugs acting on the calcium receptor. Calcimimimetics and calcilytics. In: Bilezikian JP, Martin TJ, Raisz LG, editors. *Principles of Bone Biology*. 3rd ed San Diego: Academic Press; 2008. p. 1711–35.
290. Timmers HJLM, Karperien M, Hamdy NAT, De Boer H, Hermus ARMM. Normalization of serum calcium by cinacalcet in a patient with hypercalcaemia due to a *de novo* inactivating mutation of the calcium-sensing receptor. *J Int Med* 2006;**260**:177–82.
291. Festen-Spanjer B, Haring CM, Koster JB, Mudde AH. Correction of hypercalcemia by cinacalcet in familial hypocalciuric hypercalcemia. *Clin Endocrinol* 2007;**68**:321–5.
292. Aida K, Koishi S, Inoue M, Nakazato M, Tawata M, Onaya T. Familial hypocalciuric hypercalcemia associated with mutation in the human Ca^{2+}-sensing receptor gene. *J Clin Endocrinol Metab* 1995;**80**:2594–8.
293. Alvarez-Hernandez D, Santamaria I, Rodriguez-Garcia M, Iglesias P, Delgado-Lillo R, Cannata-Andia JB. A novel mutation in the calcium-sensing receptor responsible for autosomal dominant hypocalcemia in a family with uncommon parathyroid hormone polymorphisms. *J Mol Endocrinol* 2003;**31**:255–62.
294. Arunchaiya S, Pollak MR, Seidman CA, Pinhas-Hamiel O, Welch TR, Tsang RC, et al. Marked hypercalcemia in a five month old male associated with heterozygous point mutation in the calcium-sensing receptor gene. *Prog & Abs 80th Ann Meet Endo Soc* 1998;493 Abs. P3-535.
295. Baron J, Winer KK, Yanovski JA, Cunningham AW, Laue L, Zimmerman D, et al. Mutations in the Ca^{2+}-sensing receptor gene cause autosomal dominant and sporadic hypoparathyroidism. *Hum Mol Genet* 1996;**5**:601–6.
296. Burren CP, Curley A, Christie P, Rodda CP, Thakker RV. A family with autosomal dominant hypocalcaemia with hypercalciuria (ADHH): mutational analysis, phenotypic variability and treatment challenges. *J Pediatr Endocrinol Metab* 2005;**18**:689–99.
297. Burski K, Torjussen B, Paulsen AQ, Boman H, Bollerslev J. Parathyroid adenoma in a subject with familial hypocalciuric hypercalcemia: coincidence or causality? *J Clin Endocrinol Metab* 2002;**87**:1015–6.
298. Chikatsu N, Fukumoto S, Suzawa M, Tanaka Y, Takeuchi Y, Takeda S, et al. An adult patient with severe hypercalcaemia and hypocalciuria due to a novel homozygous inactivating mutation of calcium-sensing receptor. *Clin Endocrinol* 1999;**50**:537–43.
299. Chikatsu N, Watanabe S, Takeuchi Y, Muraosa Y, Sasaki S, Oka Y, et al. A family of autosomal dominant hypocalcemia with an activating mutation of calcium-sensing receptor gene. *Endocrine J* 2003;**50**:91–6.

300. Chou YH, Pollak MR, Brandi ML, Toss G, Arnqvist H, Atkinson AB, et al. Mutations of the human Ca(2+)-sensing receptor gene that cause familial hypocalciuric hypercalcemia. *Am J Hum Genet* 1995;**56**:1075–9.
301. Christie PT, Curley AJ, Harding B, Bowl MR, Turner JJO, Cappucco FP, et al. An activating calcium sensing receptor mutation associated with normocalcemic (idiopathic) hypercalciuric nephrolithiasis. *J Bone Miner Res* 2002;**17s1**:S127 Abs 1008.
302. Christie P, Curley A, Nesbit MA, Harding B, Bowl M, Thakker R. Characterization of 25 calcium-sensing receptor mutations in disorders of calcium homeostasis. In: Society for Endocrinology BES 2007 Endocrine Abstracts 13, P1.
303. Cole DR, Soule S, Raizis A, George P. Hypocalcemia and a novel calcium-sensing receptor activating mutation. *Ann Sci Meet Royal Australasian College Physicians 2005 Int Med J* (Suppl.) 2005;**35**:A68.
304. Conley YP, Finegold DN, Peters DG, Cook JS, Oppenheim DS, Ferrell RE. Three novel activating mutations in the calcium-sensing receptor for autosomal dominant hypocalcemia. *Mol Genet Metab* 2000;**71**:591–8.
305. Defrance-Faivre F, Odou M-F, Porchet N, Weill J, Guedj A, Cardot-Bauters C, et al. Molecular analysis of the calcium-sensing receptor (CaSR) gene in 40 patients suspected to have familial hypocalciuric hypercalcemia (FHH). In: European Congress of Endocrinology 2008 Endocrine Abstracts 16, OC4.7.
306. De Luca F, Ray K, Mancilla EE, Fan GF, Winer KK, Gore P, et al. Sporadic hypoparathyroidism caused by *de novo* gain-of-function mutations of the Ca(2+)-sensing receptor. *J Clin Endocrinol Metab* 1997;**82**:2710–5.
307. Demedts M, Lissens W, Wuyts W, Matthus G, Thomeer M, Bouillon R. A new missense mutation in the *CASR* gene in familial interstitial lung disease with hypocalciuric hypercalcemia and defective granulocyte function. *Am J Resp Crit Care Med* 2008;**177**:558–9.
308. Despert F, Lienhardt-Roussie A, Lardy H, Payen V, Suc AL, Magdelaine C, et al. Severe neonatal hyperparathyroidism due to a large homozygous deletion of the calcium-sensing receptor gene requiring peritoneal dialysis before surgical care. In: Prog Abs 87th Ann Mtg Endo Soc 2005 Abs P2-607.
309. Dreimane D, Hendy GN, Alon U, Geffner M. Normalization of serum calcium, phosphorus, and magnesium with homeopathic PTH in a child with hypocalcemic hypercalciuria (HCHC) and a mutation of the calcium-sensing receptor gene. In: Prog Abs 83rd Ann Mtg Endo Soc 2001 Abs P3-125.
310. D'Souza-Li L, Canaff L, Janicic N, Cole DEC, Hendy GN. An acceptor slice site mutation in the calcium-sensing receptor (CASR) gene in familial hypocalciuric hypercalcemia and neonatal severe hyperparathyroidism. *Hum Mutat* 2001;**18**:411–21.
311. D'Souza-Li L, Silva MBD, Grande MT, Carvalho D, Guerra G. Two novel mutations in the calcium-sensing receptor gene in familial hypocalciuric hypercalcemia and neonatal severe hyperparathyroidism. *J Bone Miner Res* 2002;**17s1**:M438.
312. Ekhzaimy A, Wing SS, Hendy GN. A novel mutation in the calcium-sensing receptor (CASR) gene in a hypocalcemic patient presenting with symptoms of lupus. In: CDA/CSEM Professional Conference and Annual Meetings October 18–21, 2006 Abstract.
313. Felderbauer P, Hoffman P, Klein W, Bulut K, Ansorge N, Epplen JT, et al. Identification of a novel calcium-sensing receptor gene mutation causing familial hypocalciuric hypercalcemia by single-strand conformation polymorphism analysis. *Exp Clin Endocrinol Diabetes* 2005;**113**:31–4.
314. Felderbauer P, Klein W, Bulut K, Ansorge N, Dekomien G, Werner I, et al. Mutations in the calcium-sensing receptor: a new genetic risk factor for chronic pancreatitis? *Scand J Gastroenterol* 2006;**41**:343–8.
315. Fox L, Sadowsky J, Pringle KP, Kidd A, Murdoch J, Cole DEC, et al. Neonatal hyperparathyroidism and pamidronate therapy in an extremely premature infant. *Pediatrics* 2007;**120**: e1350–4.

316. Fukumoto S, Chikatsu N, Okazaki R, Takeuchi Y, Tamura Y, Murakami T, et al. Inactivating mutations of calcium-sensing receptor results in parathyroid lipohyperplasia. *Diagn Mol Pathol* 2001;**10**:242–7.
317. Haag C, Schulze E, Lorenz A, Frank-Raue K, Raue F. Novel mutations of the calcium-sensing receptor in familial hypocalciuric hypercalcemia and autosomal dominant hypocalcemia. In: Prog & Abs 87th Ann Meet Endo Soc 2005 P2-407.
318. Hannan F, Nesbit MA, Christie P, Bex M, Bouillon R, Thakker R. A novel homozygous inactivating mutation, Pro339Thr, of the calcium-sensing receptor is associated with isolated primary hyperparathyroidism. In: Endo Abstracts 2007 13, OC9.
319. Hendy GN, Minutti C, Canaff L, Pidasheva S, Yang B, Nouhi Z, et al. Recurrent familial hypocalcemia due to germline mosaicism for an activating mutation of the calcium-sensing receptor gene. *J Clin Endocrinol Metab* 2003;**88**:3674–81.
320. Henn M, Wygoda S, Nagel M, Richter T. Benigne familiare hypokalziurische hyperkalzamie (FHH)—neue mutation des calcium sensing receptors (CaSR) bei einem 7-juhrigen madchen. In: Jahrestagung der Sachsisch-Thuringischen Gesellschaft fur Kinder- und Jugendmedizin und Kinderchirurgie am 04. und 05. April 2008 in Chenitz Abstract P29.
321. Hirai H, Nakajima S, Miyauchi A, Nishimura K, Shimizu N, Shima M, et al. A novel activating mutation (C129S) in the calcium-sensing receptor in a Japanese family with autosomal dominant hypocalcemia. *J Hum Genet* 2001;**46**:41–4.
322. Hu J, Mora S, Colussi G, Proverbio MC, Jones K, Bolzoni L, et al. Autosomal dominant hypocalcemia caused by a novel mutation in the loop 2 region of the human calcium receptor extracellular domain. *J Bone Miner Res* 2002;**17**:1461–9.
323. Hu J, Mora S, Weber G, Zamproni I, Proverbio MC, Spiegel AM. Autosomal dominant hypocalcemia in monozygotic twins caused by a *de novo* germline mutation near the amino-terminus of the human calcium receptor. *J Bone Miner Res* 2004;**19**:578–86.
324. Inoue D, Saika M, Ikeda Y, Matsumoto T. Successful treatment of hypoparathyroidism caused by a novel calcium-sensing receptor mutation with thiazide diuretics and low dose alfacalcidol. *Bone* 1998;**23S**:S382.
325. Jap T-S, Wu YC, Jenq SF, Won GS. A novel mutation in the calcium-sensing receptor gene in a Chinese subject with persistent hypercalcemia and hypocalciuria. *J Clin Endocrinol Metab* 2001;**86**:13–5.
326. Kapoor A, Satishchandra P, Ratnapriya R, Reddy R, Kadandale J, Shankar SK, et al. An idiopathic epilepsy syndrome linked to 3q13.3–q21 and missense mutations in the extracellular calcium sensing receptor gene. *Ann Neurol* 2008;**64**:158–67.
327. Lam C-W, Lee K-F, Chan AOK, Poon PMK, Law T-K, Tong S-F. Novel missense mutation in the *CASR* gene in a Chinese family with familial hypocalciuric hypercalcemia. *Clin Chim Acta* 2005;**360**:167–72.
328. Leech C, Lohse P, Stanojevic V, Lechner A, Goke B, Spitzweg C. Identification of a novel inactivating R465Q mutation of the calcium-sensing receptor. *Biochem Biophys Res Commun* 2006;**342**:996–1002.
329. Lienhardt A, Bai M, Lagarde J, Kottler M, Farhid NR. New mutation of the calcium-sensing receptor gene associated with isolated hypoparathyroidism. In: Prog & Abs 83rd Ann Meet Endo Soc 2001 P3-123.
330. Lienhardt A, Garabedian M, Bai M, Sinding C, Zhang Z, Lagarde JP, et al. A large homozygous or heterozygous in-frame deletion within the calcium-sensing receptor's carboxyl-terminal cytoplasmic tail that causes autosomal dominant hypocalcemia. *J Clin Endocrinol Metab* 2000;**85**:1695–702.
331. Lienhardt A, Young J, Largarde J, Bai M, Kottler M, Schalson G, et al. New type of calcium-sensing receptor mutation leading to isolated hypoparathyroidism. In: Prog & Abstr Endo Soc 83rd Ann Meet 2001 P1-504.

332. Livadariu E, Rydlewski C, Hamoir E, Betea D, Burlacu C, Daly AF, et al. Two novel mutations of the calcium sensing receptor gene. In: Prog & Abs Endo Soc 90th Ann Meet 2008 P1-568.
333. Lovlie R, Eiken HG, Sorheim JI, Boman H. The Ca^{2+}-sensing receptor gene (PCAR1) mutation T151M in isolated autosomal dominant hypoparathyroidism. *Hum Genet* 1996;**98**:129–33.
334. Ma RC, Lam CW, So WY, Tong PC, Cockram CS, Chow CC. A novel CASR gene mutation in an octogenarian with asymptomatic hypercalcemia. *Hong Kong Med J* 2008;**14**:226–8.
335. Mahto R, Tahraru A, MacLeod A, Thakker RV. Familial hypocalciuric hypercalcemia (FHH) caused by P748L mutation in the calcium sensing receptor (CaSR) gene. In: European Congress of Endocrinology 2006 Endocrine Abstracts 11 P163.
336. Mancilla EE, De Luca F, Ray K, Winer KK, Fan GF, Baron J. A Ca(2+)-sensing receptor mutation causes hypoparathyroidism by increasing receptor sensitivity to Ca2+ and maximal signal transduction. *Pediatr Res* 1997;**42**:443–7.
337. Marcocci C, Borsari S, Pardi E, Dipollina G, Giacomelli T, Pinchera A, et al. Familial hypocalciuric hypercalcemia in a woman with metastatic breast cancer: a case report of mistaken identity. *J Clin Endocrinol Metab* 2003;**88**:5132–6.
338. Mittelman SD, Hendy GN, Fefferman RA, Canaff L, Mosesova I, Cole DEC, et al. A hypocalcemic child with a novel activating mutation of the calcium-sensing receptor gene: successful treatment with recombinant human parathyroid hormone. *J Clin Endocrinol Metab* 2006;**91**:2474–9.
339. Miyashiro K, Kasamatsu TS, Steinmetz L, de Menezes Filho HC, Damiani D, Setian N, et al. Identification and functional analysis of a novel inactivating mutation (A804D) of the calcium-sensing receptor gene. *Clin Endocrinol* 2004;**61**:780–2.
340. Mora S, Zamproni I, Proverbio MC, Bozzetti V, Chiumello G, Weber G. Severe hypocalcemia due to a *de novo* mutation in the fifth transmembrane domain of the calcium-sensing receptor. *Am J Med Genet* 2006;**140A**:98–101.
341. Murugaian EE, Premkumar RM, Radhakrishnan L, Vallath B. Novel mutations in the calcium-sensing receptor gene in tropical chronic pancreatitis in India. *Scand J Gastroenterol* 2008;**43**:117–21.
342. Nakae J, Shinohara N, Tanahashi Y, Murashita M, Abe S, Hasegawa Y, et al. New mutations of calcium-sensing receptor gene in two Japanese patients with sporadic hypoparathyroidism with hypercalciuria. In: XVth International Symposium of Endocrinology and Development Horm. Res 48, 1997 Abstract 798, Paris, France.
343. Nakayama T, Minato M, Nakagawa M, Soma M, Tobe H, Aoi N, et al. A novel mutation in Ca^{2+}-sensing receptor gene in familial hypocalciuric hypercalcemia. *Endocrine* 2001;**15**:277–82.
344. Nissen PH, Christensen SE, Heickendorff L, Brixen K, Moekilde L. Molecular genetic analysis of the calcium sensing receptor gene in patients clinically suspected to have familial hypocalciuric hypercalcemia: phenotypic variation and mutation spectrum in a Danish population. *J Clin Endocrinol Metab* 2007;**92**:4373–9.
345. Nyweide K, Feldman KW, Gunther DF, Done S, Lewis C, Van Eenwyk C. Hypocalciuric hypercalcemia presenting as neonatal rib fractures. *Pediatr Emerg Care* 2006;**22**:722–4.
346. Obermannova B, Banghova K, Sumnik Z, Dvorakova HM, Betka J, Fenci F, et al. Unusually severe phenotype of neonatal primary hyperparathyroidism due to a heterozygous inactivating mutation in the CASR gene. *Eur J Pediatr* 2009;**168**:569–73.
347. Okazaki R, Chikatsu N, Nakatsu M, Takeuchi Y, Alima M, Miki J, et al. A novel activating mutation in calcium-sensing receptor gene associated with a family of autosomal dominant hypocalcemia. *J Clin Endocrinol Metab* 1999;**84**:363–6.
348. Ono Y, Oda N, Ishihara S, Shimomura A, Hayakawa N, Suzuki A, et al. Insulinoma cell calcium-sensing receptor influences insulin secretion in a case with concurrent familial hypocalciuric hypercalcemia and malignant metastatic insulinoma. *Eur J Endocrinol* 2008;**159**:81–9.

349. Poppe K, Karmali R, Lissens W, Vanhaelst L, Liebaers I, Fuss M, et al. Citrate infusion test in the diagnosis of hypocalcemia due to a mutation in the calcium-sensing receptor gene. *Eur J Intern Med* 2002;**13**:276–9.
350. Rajguru M, Bedu A, Magdelaine C, Bai M, Aujuard Y, Lienhardt A. Neonatal primary hyperparathyroidism due to a new calcium sensing receptor mutation. *Pediatr Res* 2001;**49**: P3–294.
351. Ryan J, Thorne J, Hoashi S, Green A, Powell D. A novel calcium-sensing receptor gene mutation in a family with an extensive history of familial hypocalciuric hypercalcemia. In: British Endocrine Societies Joint Meeting 2005 Endocrine Abstracts P, P183.
352. Sanda S, Schlingmann KP, Newfield RS. Autosomal dominant hypoparathyroidism with severe hypomagnesemia and hypocalcemia, successfully treated with recombinant PTH and continuous magnesium infusion. *J Pediatr Endocrinol Metab* 2008;**21**:385–91.
353. Sarli M, Fradinger E, Zanchetta J. Hypocalciuric hypercalcemia due to *de novo* mutation of the calcium-sensing receptor. *Medicina (Buenos Aires)* 2004;**64**:337–9.
354. Sato K, Hasegawa Y, Nakae J, Nanao K, Takahashi I, Tajima T, et al. Hydrochlorothiazide effectively reduces urinary calcium excretion in two Japanese patients with gain-of-function mutations of the calcium-sensing receptor gene. *J Clin Endocrinol Metab* 2002;**87**:3068–73.
355. Schwarz P, Larsen NE, Lonborg Friis IM, Lillquist K, Brown EM, Gammeltoft S. Familial hypocalciuric hypercalcemia and neonatal severe hyperparathyroidism associated with mutations in the human Ca^{2+}-sensing receptor gene in three Danish families. *Scand J Clin Lab Invest* 2000;**60**:221–7.
356. Shiohara M, Mori T, Bai M, Brown EM, Watanabe T, Yasuda T. A novel gain-of-function mutation (F821L) in the transmembrane domain of calcium-sensing receptor is a cause of severe sporadic hypoparathyroidism. *Eur J Pediatr* 2004;**163**:94–8.
357. Sinha SK, Ten S, Hendy GN. Novel inactivating mutation of the calcium-sensing receptor associated with familial hypocalciuric hypercalcemia. In: Pediatric Academic Societies Annual Meeting. Honolulu Hawaii May 3–6, 2008 Abstract.
358. Speer G, Toth M, Niller H-H, Salamon D, Takacs I, Miheller P, et al. Calcium metabolism and endocrine functions in a family with familial hypocalciuric hypercalcemia. *Exp Clin Endocrinol Diabetes* 2003;**111**:486–90.
359. Stock JL, Brown RS, Baron J, Coderre JA, Mancilla E, De Luca F, et al. Autosomal dominant hypoparathyroidism associated with short stature and premature osteoarthritis. *J Clin Endocrinol Metab* 1999;**84**:3036–40.
360. Suzuki M, Aso T, Sato T, Michimata M, Kazama I, Saiki H, et al. A case of gain-of-function mutation in calcium-sensing receptor: supplemental hydration is required for renal protection. *Clin Nephrol* 2005;**63**:481–6.
361. Tan YM, Cardinal J, Franks AH, Mun HC, Lewis N, Harris LR, et al. Autosomal dominant hypocalcemia: a novel activating mutation (E604K) in the cysteine-rich domain of the calcium-sensing receptor. *J Clin Endocrinol Metab* 2003;**88**:605–10.
362. Toke J, Czirjak G, Patocs A, Enyedi B, Gergics P, Csakvary V, et al. Neonatal severe hyperparathyroidism associated with a novel *de novo* heterozygous R551K inactivating mutation and a heterozygous A986S polymorphism of the calcium-sensing receptor gene. *Clin Endocrinol* 2007;**67**:385–92.
363. Uckun-Kitapci A, Underwood LE, Zhang J, Moats-Staats B. A novel mutation (E767K) in the second extracellular loop of the calcium-sensing receptor in a family with autosomal dominant hypocalcemia. *Am J Med Genet* 2005;**132A**:125–9.
364. Vigouroux C, Bourut C, Guerci B, Ziegler O, Magre J, Capeau J, et al. A new missense mutation in the calcium-sensing receptor in familial benign hypercalcemia associated with lipoatrophy and insulin resistant diabetes. *Clin Endocrinol* 2000;**53**:393–8.

365. Waller S, Kurzawinski T, Spitz L, Thakker R, Cranston T, Pearce S, et al. Neonatal severe hyperparathyroidism: genotype/phenotype correlation and the use of pamidronate as rescue therapy. *Eur J Pediatr* 2004;**163**:589–94.
366. Ward BK, Cameron FJ, Magno AL, McDonnell CM, Stuckey BGA, Ratajczak T. A novel homozygous deletion in the calcium-sensing receptor ligand-binding domain associated with neonatal severe hyperparathyroidism. *J Pediatr Endocrinol Metab* 2006;**19**:93–100.
367. Ward BK, Magno AL, Davis EA, Hanyaloglu AC, Stuckey BGA, Burrows M, et al. Functional deletion of the calcium-sensing receptor in a case of neonatal severe hyperparathyroidism. *J Clin Endocrinol Metab* 2004;**89**:3721–30.
368. Ward BK, Magno AL, Blitvitch BJ, Rea AJ, Stuckey BGA, Walsh JP, et al. Novel mutations in the calcium-sensing receptor gene associated with biochemical and function differences in familial hypocalciuric hypercalcemia. *Clin Endocrinol* 2006;**64**:580–7.
369. Ward BK, Stuckey BGA, Gutteridge DH, Laing NG, Pullan PT, Ratajczak T. A novel mutation (L174R) in the Ca^{2+}-sensing receptor sensing receptor gene associated with familial hypocalciuric hypercalcemia. *Hum Mutat* 1997;**10**:233–5.
370. Watanabe T, Bai M, Lane CR, Matsumoto S, Minamitani K, Minagawa M, et al. Familial hypoparathyroidism: identification of a novel gain of function mutation in transmembrane domain 5 of the calcium-sensing receptor. *J Clin Endocrinol Metab* 1998;**83**:2497–502.
371. Woo SI, Song H, Song KE, Kim DJ, Lee KW, Kim SJ, et al. A case report of familial benign hypocalciuric hypercalcemia: a mutation in the calcium-sensing receptor gene. *Yonsei Med J* 2006;**47**:255–8.
372. Yamamoto M, Akatsu T, Nagase T, Ogata E. Comparison of hypocalcemic hypercalciuria between patients with idiopathic hypoparathyroidism and those with gain-of-function mutations in the calcium-sensing receptor: is it possible to differentiate the two disorders? *J Clin Endocrinol Metab* 2000;**85**:4583–91.
373. Yamauchi M, Sugimoto T, Yamaguchi T, Yano S, Wang J, Bai M, et al. Familial hypocalciuric hypercalcemia caused by a R648stop mutation in the calcium sensing receptor gene. *J Bone Miner Res* 2002;**17**:2174–82.
374. Zhao XM, Hauache O, Goldsmith PK, Collins R, Spiegel AM. A missense mutation in the seventh transmembrane domain constitutively activates the human Ca^{2+} receptor. *FEBS Lett* 1999;**448**:180–4.
375. Den Dunnen JT, Antonarakis SE. Nomenclature for the description of human sequence variations. *Hum Genet* 2001;**109**:121–4.

Diseases Associated with Mutations of the Human Lutropin Receptor

Deborah L. Segaloff

Department of Molecular Physiology and Biophysics, The Roy J. and Lucille A. Carver College of Medicine, The University of Iowa, Iowa City, Iowa 52242

I. Introduction	97
II. The LHCGR and Human Physiology	98
III. The LHCGR Protein and the *LHCGR* Gene	99
IV. Activating Mutations of the LHCGR	101
V. Inactivating Mutations of the *LHCGR*	103
References	108

The human lutropin receptor (LHCGR) plays an integral role in male and female reproductive physiology. In response to either placental hCG or pituitary LH, gonadal LHCGR mediates its effects primarily through Gs activation. Heterozygous mutations leading to constitutive activation of the LHCGR cause gonadotropin-independent precocious puberty in males, but have no detectable effects on prepubertal or postpubertal females. Homozygous or compound heterozygous inactivating mutations of the *LHCGR* cause gonadal resistance to hCG and LH, where the clinical phenotypes associated with these mutations are closely correlated with the severity of the mutation. Inactivating mutations in 46, XY individuals cause Leydig cell hypoplasia and impairments in the differentiation of male external genitalia, the development of secondary sexual characteristics and sperm production. 46,XX siblings with inactivating LHCGR mutations exhibit infertility and varying degrees of menstrual irregularities.

I. Introduction

In the past several years, there have been several reviews written on the human lutropin receptor (LHCGR)[1,2] and naturally occurring mutations of the *LHCGR*.[3–6] In addition, some excellent web-based repositories listing LHCGR mutations have been developed. These include the sequence–structure–function analysis–GPHR resource at http://www.ssf-gphr.de;[7] the Glycoprotein-Hormone Receptors Information Systems database at http:wwwgris.ulb.ac.be;[8] and the GPCR Natural Variants Database at http://nava.liacs.nl. One of these (http://gris.ulb.ac.be/) also contains structural information on the LHCGR and mutants thereof. This review is intended to provide a brief, general overview of LHCGR, an update on naturally occurring mutations of the *LHCGR*, the physiological consequences of such mutations, and current views on the mechanisms underlying the activity and hormone responsiveness of cells expressing LHCGR mutants.

II. The LHCGR and Human Physiology

The LHCGR is essential for normal reproductive functioning in males and females. Although there are numerous reports of extragonadal expression of the LHCGR,[1,9] the primary sites of LHCGR expression are gonadal, with the LHCGR being present on testicular Leydig cells and on ovarian theca cells and differentiated granulosa cells. In males, the LHCGR plays an important role in fetal development. During this time, human chorionic gonadotropin (hCG, a glycoprotein hormone that is nearly identical to LH) is secreted by the mother's placenta and stimulates the LHCGR present on fetal Leydig cells, causing increased androgen synthesis. The fetal androgens mediate the differentiation of external genitalia into the male phenotype and stimulate descent of the testes into the scrotum. After birth and prior to puberty, the levels of LH secreted by the anterior pituitary remain at relatively low prepubertal levels. By mechanisms not yet well understood, increased pulsatile secretion of LH is initiated at the time of puberty. In postpubertal males, LH stimulates the synthesis of testosterone by testicular mature Leydig cells. The increased levels of androgens that commence with puberty stimulate the maturation of the male reproductive tract and mediate the secondary sexual characteristics that are acquired during puberty. Testosterone is also essential for spermatogenesis in the male.

In contrast to males, females do not require the LHCGR for normal fetal development. In postpubertal females (until the time of menopause), LH stimulates androgen synthesis in ovarian theca cells. The androgens serve as a substrate for FSH-stimulated estrogen production in ovarian granulosa cells. Estrogens stimulate maturation of the female reproductive tract and mediate the development of female secondary sexual characteristics, including breast development. In contrast to ovarian theca cells, which constitutively express the LHCGR, granulosa cells only express LHCGR at the latter stages of follicular development when the granulosa cells, under the actions of FSH and estradiol, have become fully differentiated and expression of the LHCGR has been induced. These mature granulosa cells respond to LH with increased progesterone synthesis. Importantly, the induction of LHCGR during granulosa cell differentiation allows the preovulatory follicle to respond to the mid-cycle surge of LH with ovulation and the release of the mature ovum. After ovulation, the ruptured follicle differentiates into the corpus luteum, which expresses LHCGR and secretes estrogen and progesterone. In the absence of fertilization of the ovum, the corpus luteum undergoes programmed cell death about 14 days after ovulation. However, if fertilization ensues, hCG secreted by the placenta "rescues" the corpus luteum, allowing the continued production of estrogen and progesterone, which are necessary for the maintenance of pregnancy.

III. The LHCGR Protein and the *LHCGR* Gene

The LHCGR is a member of the Family A or rhodopsin-like family of GPCRs. As such, it contains the canonical serpentine region composed of seven transmembrane (TM) helices interconnected by three extracellular and three intracellular loops. In addition, though, it contains a relatively large N-terminal extracellular domain (ECD) that mediates the high affinity binding of pituitary LH or placental hCG.[10] As initially translated, the LHCGR contains a signal peptide that is subsequently cleaved from the mature protein. Unlike the rat receptor, where the N terminus of the mature protein was determined experimentally,[11] the N terminus of the mature human LHCGR has not been experimentally confirmed and can only be theoretically predicted. Therefore, the numbering of the human LHCGR typically designates the first amino acid of the translated protein (i.e., first residue of the signal peptide) as residue 1.

The ECD of the LHCGR is composed largely of a series of leucine-rich repeats (LRRs).[1,12] It is flanked at one end by an N-terminal cysteine-rich domain and at the other end by another cysteine-rich domain. The latter is commonly called the hinge region because it links the LRR region to the serpentine domain. As with other GPCRs, activation of the LHCGR stabilizes the serpentine domain in an active conformation, enabling it to engage with and activate the appropriate heterotrimeric G protein. (Active and inactive states of the LHCGR each most likely represents an ensemble of conformations,[13–15] but for simplicity they are referred to herein in the singular.) The primary G protein activated by agonist-occupied LHCGR or LHCGR constitutively active mutants (CAMs) is Gs, which elicits an increase in intracellular cAMP and a subsequent stimulation of steroidogenesis in gonadal cells. Activation of $G_{q/11}$ also occurs, resulting in a stimulation of the inositol phosphate pathway. However, high densities of receptor are necessary to observe stimulation of this signaling pathway.[16–18] The activation of the inositol phosphate pathway by the LHCGR in granulosa cells appears to be involved in antagonizing the induction of aromatase activity mediated by low concentrations of cAMP and may be physiologically relevant during the ovarian periovulatory and luteal phases.[16–18] Although the $G_{i/o}$ family of G proteins has also been shown to be stimulated with LHCGR activation, the functional consequences of this are not known.[19] How the binding of LH or hCG to the ECD stabilizes the serpentine domain in an active conformation is not well understood and is an area of active investigation. In this context, it is also worthwhile to consider the other glycoprotein-hormone receptors, the human follitropin receptor (hFSHR), and human thyrotropin receptor (hTSHR), which share extensive homology to the LHCGR and possess the same overall structural organization. Studies on all three glycoprotein-hormone receptors suggest that the hinge region may play a key role in activation of the receptor.[20–22] Although the structures of the serpentine regions of the glycoprotein-hormone receptors have not been determined, one can use the structures of rhodopsin or other Family A GPCRs whose crystal structures have been determined[23–30] as the basis for homology models of the TM helices of the LHCGR. With respect to the LHCGR ECD, Fan and Hendrickson recently determined the structure of the hormone-binding domain of the hFSHR (hFSHR$_{HB}$) complexed with FSH.[12,31] In this study, the hinge region of the hFSHR ECD was not included because it impaired crystallization. These studies elegantly demonstrated that the LLRs of the ECD form a tube-like structure bent into a horseshoe structure, with FSH bound to the inner concave surface as if in a handclasp. Given the high degree of homology between the hFSHR and the LHCGR and between FSH and LH or hCG, it is possible to fairly accurately predict the tertiary structure of the hormone-binding area of the LHCGR complexed with LH or hCG. However, because the hFSHR$_{HB}$ construct used in the structural studies lacked the hinge region, it is not yet possible to accurately predict the juxtaposition of the ECD of the LHCGR (or any glycoprotein-hormone receptor) relative to its serpentine domain.

The gene encoding the LHCGR (*LHCGR*; NCBI Gene ID 3973; http://www.ncbi.nlm.nih.gov/) is located on the short arm of chromosome 2 (2p21).[32] It spans ~80 kb and has been thought to be composed of 11 exons and 10 introns. Exon 11 of the gene encodes the entire serpentine domain as well as the C-terminal portion of the hinge region. The N-terminal portion of the hinge region is encoded by exon 10 and the signal peptide and remaining portion of the ECD are encoded by exons 1–9. Recently, a novel primate-specific exon (termed exon 6A) has been identified within intron 6 of the *LHCGR* gene.[33] This exon is not used by the wild-type (wt) full-length receptor. When exon 6A is utilized, it results in a truncated LHCGR protein of only 209 amino acids that remains trapped intracellularly. However, it is thought that mRNA transcripts arising from the usage of exon 6A would also be targeted for nonsense-mediated mRNA decay, a cellular process that degrades nonsense mRNAs.[33]

IV. Activating Mutations of the LHCGR

In 1993, Shenker *et al.* reported the first constitutively activating mutation of the LHCGR in a young boy with familial male-limited gonadotropin-independent precocious puberty and Leydig cell hyperplasia (also referred to as testotoxicosis).[34] An activating mutation of the LHCGR was suspected because precocious puberty and elevated levels of androgens were observed in spite of prepubertal levels of LH. The boy was found to have a heterozygous mutation of the *LHCGR*, causing a D578G substitution within TM6 of the LHCGR. Heterologous cells transfected with a cDNA encoding the mutant form of the LHCGR exhibited a small, but significant, increase in basal cAMP compared to cells transfected with a cDNA encoding the wt form of the receptor, confirming that the mutant LHCGR was constitutively active. Although the increase in basal cAMP levels caused by the mutant LHCGR *in vitro* was fairly modest, it was clearly sufficient within a physiological context to stimulate androgen synthesis and, therefore, cause precocious puberty. Since then, many other activating mutations of the *LHCGR* have been identified (summarized in Table I) and it has been determined that the D578G mutation is the most common activating *LHCGR* mutation in the United States.[35]

TABLE I
NATURALLY OCCURRING ACTIVATING MUTATIONS OF THE LHCGR

LHCGR mutations	Locations	References
$L^{368}P$	TM1	40
$A^{373}V$	TM1	68
$M^{398}T$	TM2	35, 69–72
$L^{457}R$	TM3	42, 43
$I^{542}L$	TM5	35
$D^{564}G$	TM6	35, 73
$A^{568}V$	TM6	74
$M^{571}I$	TM6	75, 76
$A^{572}V$	TM6	77
$I^{575}L$	TM6	73, 78
$T^{577}I$	TM6	76, 79
$D^{578}G$	TM6	34, 35, 75, 80, 81
$D^{578}Y$	TM6	35, 71, 82
$D^{578}E$	TM6	83
$D^{578}H$	TM6	36, 37, 84
$Cys^{581}R$	TM6	35

Of all the naturally occurring activating *LHCGR* mutations identified in boys with gonadotropin-independent precocious puberty, only one, D578H, has been found in boys also exhibiting Leydig cell adenomas.[36–38] It was originally reported that D578H, when expressed in heterologous cells, preferentially activated the inositol phosphate signaling pathway.[36] Since then, more extensive studies have been performed examining the signaling pathways activated by D578H and by other LHCGR CAMs. These studies have shown there to be no difference between D578H and other LHCGR CAMs with respect to the G proteins activated or the second messenger pathways stimulated.[16,19] The one salient feature that sets D578H apart from the other CAMs, however, is that D578H is the only *LHCGR* activating mutation thus far to have been identified as a somatic mutation.[36–38] All the other *LHCGR* activating mutations have been familial or sporadic germline mutations.

It should be noted that activating mutations of the *LHCGR* have only been associated with gonadotropin-independent precocious puberty in boys, not girls. Female siblings or mothers of affected boys who also harbored heterozygous activating *LHCGR* mutations did not undergo precocious puberty. This is not entirely unexpected because prepubertal females would not yet express LHCGRs on granulosa cells due to the lack of activation of the hFSHR. However, the absence of any noticeable reproductive abnormalities in postpubertal women possessing an activating mutation of the *LHCGR*, in particular hyperandrogenism, has been surprising.[39–41]

All naturally occurring activating mutations of the *LHCGR* that have been identified thus far result in amino acid substitutions within the serpentine region of the LHCGR, specifically within the predicted TM alpha helices. When introduced into heterologous cells, LHCGR CAMs cause an increase in basal cAMP relative to wt receptor. Most LHCGR CAMs exhibit basal levels of cAMP that are not as great as those observed in cells expressing the wt LHCGR that have been incubated with a maximally effective concentration of hCG, suggesting that the mutations stabilize the LHCGR in states of intermediate activation. Although most LHCGR CAMs respond further to hormonal stimulation, some (such as I542L, L457R, D578Y, and D578H) do not.[16,42–44] It is likely that in cases such as these the mutation independently increases constitutive activity while also perturbing interactions necessary for hormone-mediated stabilization of the receptor in a more active conformation. It is also worth noting that, in contrast to the hFSHR, which displays relatively strong promiscuous activation by other glycoprotein hormones when it is constitutively active,[45–49] LHCGR activating mutants show little or no promiscuous activation.[44]

In collaboration with our laboratory and others, Dr. Francesca Fanelli has used computational analyses to predict the structural changes in the TM helices accompanying LHCGR activation.[44,50–54] Because rhodopsin was the first Family A GPCR for which a high-resolution structure was available,[23] the coordinates of the rhodopsin TMs were used to create homology models for the TMs of the LHCGR. Molecular dynamic simulations were then performed on the wt LHCGR and various LHCGR CAMs, both naturally occurring and of laboratory design, to determine structural changes in the TMs common to the LHCGR CAMs and distinct from the wt LHCGR. In spite of the disparate locations of the various activating mutations, a common structural change observed in all LHCGR CAMs examined was an increase in the solvent accessible surface area of residues at the cytoplasmic ends of TMs 3 and 6.[44,50-54] This was concomitant with the disruption of charge-reinforced hydrogen bonding interactions between R464 of the E/D-R-Y/W at the cytoplasmic end of TM3 and E463 and D564 at the cytoplasmic end of TM6. These studies suggest that activation of the LHCGR is accompanied by an increase in the cytoplasmic cleft between the cytoplasmic ends of TMs 3 and 6, resulting in the exposure of residues necessary for G protein activation. These *in silico* findings are in agreement with earlier *in vitro* studies suggesting that the cytoplasmic end of TM6 of the rat LHR activates Gs[55] and with more recent structural studies on the activated form of opsin that suggest an interaction of the cytoplasmic end of TM6 with the alpha subunit of the heterotrimeric G protein.[25] Although the final conformational outcome of any given LHCGR activating mutation appears to be a disruption of the salt bridge at the cytoplasmic ends of TMs 3 and 6, the local conformational changes caused by different activating mutation are most likely distinct. The immediate structural perturbations of a mutation may involve the disruption of existing interhelical interactions and/or the introduction of new interhelical interactions. An example of the latter is the activating mutation L457R. In this case, the introduction of an arginine in place of a highly conserved leucine in the central portion of TM3 now allows for the formation of a salt bridge between R457 and D578 in TM6. The formation of this salt bridge ultimately causes conformational changes in the LHCGR that result in a disruption of the key salt bridge between the cytoplasmic ends of TMs 3 and 6.[52]

Recent studies have shown that the LHCGR is constitutively expressed as dimers and higher ordered oligomers and that these form early in the biosynthetic pathway.[56,57] The dimerization of the LHCGR does not appear to be influenced by hormone activation and CAMs of the LHCGR display the same relative propensity for homodimerization as the wt LHCGR.[57]

V. Inactivating Mutations of the *LHCGR*

Inactivating or loss-of-function mutations of the *LHCGR* cause target cells to exhibit a reduced or absent responsiveness to hCG and LH (although one mutation, as discussed in this section, shows a selective loss of responsiveness to LH). Ultimately, the phenotype resulting from an *LHCGR* inactivating mutation correlates well with the degree of resistance to hormone stimulation as determined in patients challenged with exogenous hCG or in heterologous cells expressing the mutant receptor that have been incubated with hCG or LH.[58] Unlike activating *LHCGR* mutations, which cause a phenotype when expressed in heterozygous form, inactivating *LHCGR* mutations give rise to a clinical phenotype only when expressed in homozygous or compound heterozygous form. Consequently, these are more rare. The known *LHCGR* inactivating mutations are summarized in Table II. All have been identified in individuals with a 46,XY karyotype, where the decreased sensitivity of Leydig cells to hCG and LH results in decreased testosterone production and the consequent disruption of fetal and pubertal male development, resulting in

TABLE II
NATURALLY OCCURRING INACTIVATING MUTATIONS OF THE LHCGR

LHCGR mutations	Locations	References
Insertion of LLKLLLLLQ	Signal peptide, codon 18, exon 1	67, 85
I114F	ECD, exon 4	86
C131R	ECD, exon 5	87
V144F	ECD, exon 5	88
F194V	ECD, exon 7	89
Unique truncated 209aa LHCGR	Exon 6A	33
ΔExon 8	ECD, exon 8	90
ΔExon 10	ECD, exon 10	64
ΔY317-S324	ECD, exon 11	63
C343S	ECD, exon 11	91
E354K	ECD, exon 11	92
I374T	TM1, exon 11	93
T392I	IL1, exon 11	93
W491°	TM4, exon 11	67
C543R	TM5, exon 11	91
C545°	TM5, exon 11	85, 94
R554°	IL3, exon 11	95
A589fs	TM6, exon 11	66
A593P	TM6, exon 11	58, 96, 97
ΔL608,V609	TM7, exon 11	59
S616Y	TM7, exon 11	58, 90, 95
I625K	TM7, exon 11	58, 67

Leydig cell hypoplasia (LCH). Type I LCH is used to describe the more severe form of LCH associated with *LHCGR* mutations causing a complete disruption of hCG and LH signaling. This results in a very rare form of 46,XY disorder of sexual development, characterized by a predominantly female external phenotype with a blind-ending vagina and cryptorchidism. Typically, these individuals have been raised as females and have been diagnosed in their teens or early twenties when medical treatment was sought due to lack of breast development and amenorrhea. Type II LCH is a milder form of LCH that is characterized by *LHCGR* mutations that retain some degree of responsiveness to hCG and LH. These 46,XY individuals exhibit a range of incomplete male sexual differentiation, often resulting in micropenis and/or hypospadias. 46,XX siblings of affected 46,XY individuals, who similarly carry homozygous or compound heterozygous inactivating mutations of the *LHCGR*, are infertile and exhibit primary or secondary amenorrhea. The 46,XX females with ovarian resistance to LH due to inactivating *LHCGR* mutations have normal female external genitalia and undergo normal breast development at puberty. In this respect, ovarian and adrenal androgens appear to be an adequate source for the estrogen synthesis mediating pubertal development in the affected 46,XX females. In addition, they exhibit normal or enlarged cystic ovaries with elevated LH and LH/FSH ratio, measurable estradiol levels and normal androgen levels. Therefore, while stimulation of a functional LHCGR is essential to normal ovarian function (to achieve preovulatory levels of estrogen, stimulate ovulation, induce formation of the corpus luteum, and sustain the corpus luteum through the luteal phase), it is not essential for female pubertal development. As would be expected, both genetic males and females with inactivating *LHCGR* mutations display elevated levels of serum LH as a result of insufficient negative feedback of gonadal steroid hormones on the anterior pituitary.

As summarized in Table II, most inactivating mutations of the *LHCGR* are missense mutations that result in single amino acid substitutions in the LHCGR. In addition, mutations causing amino acid deletions, amino acid insertions, or premature truncations of the receptor have also been reported. Although all of these mutations have been shown to result in little or no hormone-stimulated cAMP by heterologous cells expressing the mutant, the underlying mechanism(s) for the decreased responsiveness was not investigated in many cases. At the very least, mutations may impact the steady-state level of total (i.e., mature and immature) LHCGR expressed, the folding of the immature receptor (and consequently its exiting from the endoplasmic reticulum (ER) and trafficking to the plasma membrane), the binding affinity of the receptor, and/or the intrinsic signaling properties of the receptor. These impairments are not mutually exclusive. For example, the ΔL608,V609 mutant has been shown to result in decreased total levels of mutant LHCGR, retention of most of the mutant receptor intracellularly, and an impairment in hCG-stimulated signaling (in spite of normal hormone binding) by the small number of mutant receptors that are expressed on the cell surface.[59]

Where examined, the most prevalent molecular basis for reduced target cell responsiveness accompanying an LHCGR mutation has been the intracellular retention of the mutant receptor. This is thought to be due to a misfolding of the mutant receptor, which is detected by the cell's quality control mechanisms in the ER, causing it to be retained in the ER. The folding of the LHCGR has been shown to involve its association with several ER resident chaperone proteins.[60] Interestingly, two inactivating LHCGR mutants, S616Y and A593P, interact with a different cohort of ER chaperone proteins than the wt LHCGR.[60] Furthermore, the pattern of chaperone proteins is somewhat different between the two mutants, perhaps reflective of different stages of folding that are attained by each of the mutants. Whereas some LHCGR inactivating mutants are retained almost exclusively in the ER and consequently display little or no cell surface expression, others are only partially retained in the ER and exhibit detectable cell surface expression, albeit reduced as compared to the wt receptor. Presumably, the severity of misfolding determines the extent to which a given mutant will be retained intracellularly. Because these various inactivating mutations primarily affect the folding of the LHCGR, they are scattered throughout the receptor, being present in both the ECD as well as the serpentine region. It has been shown *in vitro* that coexpression of a misfolded inactivating LHCGR mutant with the wt LHCGR results in decreased cell surface expression of the wt receptor due to dimerization between the two forms of the receptor and the consequent partial retention of wt LHCGR in the ER.[56,57] In spite of this, a clinical phenotype is not observed in individuals who are heterozygous for LHCGR misfolded inactivating mutations. This is presumably because gonadal cells express a large number of spare LHCGRs[61,62] and the reduction in cell surface gonadal LHCGR expression in the heterozygotes (resulting from both the decreased cell surface expression of the mutant receptor as well as decreased expression of the wt receptor due to a dominant negative effect of the mutant receptor) must not be sufficient to cause a decrease in hormone responsiveness in the gonadal cell.

Although most LHCGR mutations arise due to misfolding and intracellular retention of the mutant receptor, one mutation has recently been identified in a 46,XY male with a particularly mild phenotype in which the LHCGR mutant is expressed normally at the cell surface, but it has a slightly reduced potency for LH and hCG-stimulated cAMP.[63] In this case, a missense mutation at position-1 of the intron 10-exon 11 boundary of the *LHCGR* causes the first 24 bases of exon 11 to be spliced out. This results in the deletion of eight amino acids in the C-terminal cysteine-rich hinge region of the LHCGR.[63] The relatively mild reduced potency for LH and hCG correlated with the clinical presentations.

Thus, the 46,XY affected male exhibited delayed puberty, micropenis, and oligospermia. Two of his sisters were homozygous for the same mutation and both were infertile. Although one sister had a phenotype consistent with *LHCGR* inactivating mutations, the other sister had a much milder phenotype than had been previously observed for 46,XX individuals with inactivating *LHCGR* mutations. Thus, remarkably, she exhibited normal menses for 4 years prior to the onset of amenorrhea and her LH levels were not elevated.

In recent years, other unusual inactivating mutations of the *LHCGR* have been reported that warrant specific discussions. In one instance, a mutation causing the deletion of exon 10 of the *LHCGR* was identified in a patient with Type II LCH. This 46,XY individual had a normal male phenotype, but lacked pubertal development and was hypogonadal, suggesting that the mutant LHCGR was responsive to fetal hCG (normal differentiation of the external genitalia), but refractory to pubertal LH.[64] Indeed, *in vitro* characterization of the mutant LHCGR lacking the residues encoded for by exon 10 confirmed this hypothesis.[65] Interestingly, the binding affinities for LH and hCG were normal, suggesting that exon 10 is necessary for LH, but not hCG, mediated stabilization of the LHCGR in an active conformation. In another instance, a single nucleotide insertion was identified in exon 11 (codon A589fs) that causes a frameshift error in the coding sequence, resulting in the subsequent substitution of 16 residues and then the insertion of a stop codon, causing premature truncation of the receptor.[66] Not surprisingly, this mutation (which would be expected to be completely inactive) was identified in a 46,XY child with female external genitalia and palpable testes. Although frameshift mutations are common in other genes, this was the first to be identified in the *LHCGR*.

A particularly novel *LHCGR* inactivating mutation was recently reported involving a novel, cryptic exon (termed exon 6A) located in intron 6A of the *LHCGR* that caused male pseudohermaphroditism.[33] It has been perplexing to those in the field that a number of 46,XY patients have been identified with clinical presentations consistent with Type I LCH and yet mutations in the coding sequence of the *LHCGR* could not be identified.[67] To address this question, Kossack *et al.* embarked on a study to clone *LHCGR* variants from a cynomolgus testis library and a human granulosa cell library. In doing so, they identified a primate-specific exon 6A, which can be spliced onto novel *LHCGR* transcripts. The full-length wt LHCGR does not utilize exon 6A. Indeed, if exon 6A is used, it results in truncated LHCGRs consisting of only the first 209 amino acids. Not surprisingly, expression of the truncated LHCGRs in heterologous cells indicates that they are retained intracellularly. Screening of LCH patients identified two distinct mutations in exon 6A (nucleotide base changes A557C or G558C). *In vitro* experiments showed that each of these mutations resulted in an increase in the proportion of transcripts utilizing exon 6A. Transfection of heterologous cells with minigenes encoding for the mutant

LHCGRs showed little to no response to hCG. The authors suggest that the exon 6A containing transcripts normally undergo nonsense-mediated mRNA decay, a cellular process that degrades nonsense mRNAs and that the mutations in exon 6A decrease their degradation, leading to increased expression of truncated, nonfunctional LHCGR.

Of the 16 patients with LCH screened in the above study, 13 still remain enigmatic. Therefore, there are still likely to be other mutations of the *LHCGR* that do not occur in exons 1–11 or the newly discovered exon 6A. In addition to considering mutations in the *LHCGR per se* giving rise to LCH, one can also speculate that perhaps mutations in microRNAs or proteins that regulate the levels of the *LHCGR* mRNA transcript encoding the full-length wt receptor are involved. It is also conceivable that mutations in other proteins that may specifically affect the folding, processing, cell surface targeting, or signaling properties of the LHCGR are involved. Clearly, this is an area that warrants further investigation.

Acknowledgments

I thank Drs. Mario Ascoli, Ana Claudia Latronico, Francesca Fanelli, and Terry Hébert for their collaborations, Dr. Latronico for critically reviewing the manuscript, and present and former members of my laboratory for their contributions to work discussed herein. Studies discussed from the author's laboratory were supported by NIH grants HD22196 and DK068614.

Note Added in Proof

A publication in press (Qiao, J. *et al.*, Human Mutation, DOI: 10.1002/humu.21072) reports on a patient with male pseudohermaphroditism who expresses a splicing mutation of the *LHCGR* gene and a novel mutation (I152T) that, without affecting cell surface expression of the LHCGR, causes a profound reduction in hCG binding activity.

References

1. Ascoli M, Fanelli F, Segaloff DL. The lutropin/choriogonadotropin receptor, a 2002 perspective. *Endocr Rev* 2002;**23**:141–74.
2. Vassart G, Pardo L, Costagliola S. A molecular dissection of the glycoprotein hormone receptors. *Trends Biochem Sci* 2004;**29**:119–26.
3. Latronico A, Segaloff D. Naturally occurring mutations of the luteinizing hormone receptor: lessons learned about reproductive physiology and G protein-coupled receptors. *Am J Hum Genet* 1999;**65**:949–58.
4. Themmen APN, Huhtaniemi IT. Mutations of gonadotropins and gonadotropin receptors: elucidating the physiology and pathophysiology of pituitary-gonadal function. *Endocr Rev* 2000;**21**:551–83.

5. Themmen AP. An update of the pathophysiology of human gonadotrophin subunit and receptor gene mutations and polymorphisms. *Reproduction* 2005;**130**:263–74.
6. Piersma D, Verhoef-Post M, Berns EM, Themmen AP. LH receptor gene mutations and polymorphisms: an overview. *Mol Cell Endocrinol* 2007;**260–262**:282–6.
7. Kleinau G, Brehm M, Wiedemann U, Labudde D, Leser U, Krause G. Implications for molecular mechanisms of glycoprotein hormone receptors using a new sequence–structure–function analysis resource. *Mol Endocrinol* 2007;**21**:574–80.
8. Van Durme J, Horn F, Costagliola S, Vriend G, Vassart G. GRIS: glycoprotein-hormone receptor information system. *Mol Endocrinol* 2006;**20**:2247–55.
9. Apaja PM, Harju KT, Aatsinki JT, Petaja-Repo UE, Rajaniemi HJ. Identification and structural characterization of the neuronal luteinizing hormone receptor associated with sensory systems. *J Biol Chem* 2004;**279**:1899–906.
10. Xie YB, Wang H, Segaloff DL. Extracellular domain of lutropin/choriogonadotropin receptor expressed in transfected cells binds choriogonadotropin with high affinity. *J Biol Chem* 1990;**265**:21411–4.
11. McFarland KC, Sprengel R, Phillips HS, Kohler M, Rosemblit N, Nikolics K, et al. Lutropin-choriogonadotropin receptor: an unusual member of the G protein-coupled receptor family. *Science* 1989;**245**:494–9.
12. Fan QR, Hendrickson WA. Structure of human follicle-stimulating hormone in complex with its receptor. *Nature* 2005;**433**:269–77.
13. Onaran HO, Scheer A, Cotecchia S, Costa T. A look into receptor efficacy: from the signalling network of the cell to the intramolecular motion of the receptor. In: Kenakin T, Angus JA, editors. *The Pharmacology of functional, biochemical, and recombinant receptor systems.* Berlin: Springer; 2000. p. 217–59.
14. Kenakin T. Efficacy at G-protein-coupled receptors. *Nat Rev Drug Discov* 2002;**1**:103–10.
15. Kobilka BK, Deupi X. Conformational complexity of G-protein-coupled receptors. *Trends Pharmacol Sci* 2007;**28**:397–406.
16. Hirakawa T, Galet C, Ascoli M. MA-10 cells transfected with the human lutropin/choriogonadotropin receptor (hLHR): a novel experimental paradigm to study the functional properties of the hLHR. *Endocrinology* 2002;**143**:1026–35.
17. Donadeu FX, Ascoli M. The differential effects of the gonadotropin receptors on aromatase expression in primary cultures of immature rat granulosa cells are highly dependent on the density of receptors expressed and the activation of the inositol phosphate cascade. *Endocrinology* 2005;**146**:3907–16.
18. Andric N, Ascoli M. Mutations of the lutropin/choriogonadotropin receptor that do not activate the phosphoinositide cascade allow hCG to induce aromatase expression in immature rat granulosa cells. *Mol Cell Endocrinol* 2008;**285**:62–72.
19. Hirakawa T, Ascoli M. A constitutively active somatic mutation of the human lutropin receptor found in Leydig cell tumors activates the same families of G proteins as germ line mutations associated with Leydig cell hyperplasia. *Endocrinology* 2003;**144**:3872–8.
20. Bruysters M, Verhoef-Post M, Themmen AP. Asp330 and Tyr331 in the C-terminal cysteine-rich region of the luteinizing hormone receptor are key residues in hormone-induced receptor activation. *J Biol Chem* 2008;**283**:25821–8.
21. Agrawal G, Dighe RR. Critical involvement of the hinge region of the follicle-stimulating hormone receptor in the activation of the receptor. *J Biol Chem* 2009;**284**:2636–47.
22. Mizutori Y, Chen CR, McLachlan SM, Rapoport B. The thyrotropin receptor hinge region is not simply a scaffold for the leucine-rich domain but contributes to ligand binding and signal transduction. *Mol Endocrinol* 2008;**22**:1171–82.
23. Palczewski K, Kumasaka T, Hori T, Behnke CA, Motoshima H, Fox BA, et al. Crystal structure of rhodopsin: a G protein-coupled receptor. *Science* 2000;**289**:739–45.

24. Park JH, Scheerer P, Hofmann KP, Choe HW, Ernst OP. Crystal structure of the ligand-free G-protein-coupled receptor opsin. *Nature* 2008;**454**:183–7.
25. Scheerer P, Park JH, Hildebrand PW, Kim YJ, Krauss N, Choe HW, et al. Crystal structure of opsin in its G-protein-interacting conformation. *Nature* 2008;**455**:497–502.
26. Cherezov V, Rosenbaum DM, Hanson MA, Rasmussen SG, Thian FS, Kobilka TS, et al. High-resolution crystal structure of an engineered human beta2-adrenergic G protein-coupled receptor. *Science* 2007;**318**:1258–65.
27. Rosenbaum DM, Cherezov V, Hanson MA, Rasmussen SG, Thian FS, Kobilka TS, et al. GPCR engineering yields high-resolution structural insights into beta2-adrenergic receptor function. *Science* 2007;**318**:1266–73.
28. Rasmussen SG, Choi HJ, Rosenbaum DM, Kobilka TS, Thian FS, Edwards PC, et al. Crystal structure of the human beta2 adrenergic G-protein-coupled receptor. *Nature* 2007;**450**:383–7.
29. Warne T, Serrano-Vega MJ, Baker JG, Moukhametzianov R, Edwards PC, Henderson R, et al. Structure of a beta1-adrenergic G-protein-coupled receptor. *Nature* 2008;**454**:486–91.
30. Jaakola VP, Griffith MT, Hanson MA, Cherezov V, Chien EY, Lane JR, et al. The 2.6 angstrom crystal structure of a human A2A adenosine receptor bound to an antagonist. *Science* 2008;**322**:1211–7.
31. Fan QR, Hendrickson WA. Assembly and structural characterization of an authentic complex between human follicle stimulating hormone and a hormone-binding ectodomain of its receptor. *Mol Cell Endocrinol* 2007;**260–262**:73–82.
32. Rousseau-Merck MF, Misrahi M, Atger M, Loosfelt H, Milgrom E, Berger R. Localization of the human luteinizing hormone/choriogonadotropin receptor gene (LHCGR) to chromosome 2p21. *Cytogenet Cell Genet* 1990;**54**:77–9.
33. Kossack N, Simoni M, Richter-Unruh A, Themmen AP, Gromoll J. Mutations in a novel, cryptic exon of the luteinizing hormone/chorionic gonadotropin receptor gene cause male pseudohermaphroditism. *PLoS Med* 2008;**5**:e88.
34. Shenker A, Laue L, Kosugi S, Merendino Jr JJ, Minegishi T, Cutler Jr GB. A constitutively activating mutation of the luteinizing hormone receptor in familial male precocious puberty. *Nature* 1993;**365**:652–4.
35. Laue L, Chan W-C, Hsueh A, Kudo M, Hsu SY, Wu S, et al. Genetic heterogeneity of constitutively activating mutations of the human luteinizing hormone receptor in familial male-limited precocious puberty. *Proc Natl Acad Sci USA* 1995;**92**:1906–10.
36. Liu G, Duranteau L, Carel J-C, Monroe J, Doyle DA, Shenker A. Leydig-cell tumors caused by an activating mutation of the gene encoding the luteinizing hormone receptor. *N Engl J Med* 1999;**341**:1731–6.
37. Richter-Unruh A, Wessels HT, Menken U, Bergmann M, Schmittmann-Ohters K, Schaper J, et al. Male LH-independent sexual precocity in a 3.5-year-old boy caused by a somatic activating mutation of the LH receptor in a Leydig cell tumor. *J Clin Endocrinol Metab* 2002;**87**:1052–6.
38. Kiepe D, Richter-Unruh A, Autschbach F, Kessler M, Schenk JP, Bettendorf M. Sexual pseudo-precocity caused by a somatic activating mutation of the LH receptor preceding true sexual precocity. *Horm Res* 2008;**70**:249–53.
39. Rosenthal IM, Refetoff S, Rich B, Barnes RB, Sunthornthepvarakul T, Parma J, et al. Response to challenge with gonadotropin-releasing hormone agonist in a mother and her two sons with a constitutively activating mutation of the luteinizing hormone receptor-A clinical research center study. *J Clin Endocrinol Metab* 1996;**81**:3802–6.
40. Latronico AC, Shinozaki H, Guerra Jr G, Pereira MA, Lemos Marini SH, Baptista MT, et al. Gonadotropin-independent precocious puberty due to luteinizing hormone receptor mutations in Brazilian boys: a novel constitutively activating mutation in the first transmembrane helix. *J Clin Endocrinol Metab* 2000;**85**:4799–805.

41. Latronico AC, Lins TS, Brito VN, Arnhold IJ, Mendonca BB. The effect of distinct activating mutations of the luteinizing hormone receptor gene on the pituitary-gonadal axis in both sexes. *Clin Endocrinol (Oxf)* 2000;**53**:609–13.
42. Latronico AC, Abell AN, Arnhold IJP, Liu X, Lins TSS, Brito VN, et al. A unique constitutively activating mutation in the third transmembrane helix of the luteinizing hormone receptor causes sporadic male gonadotropin independent precocious puberty. *J Clin Endocrinol Metab* 1998;**83**:2435–40.
43. Latronico AC, Segaloff DL. Insights learned from the L457(3.43)R, an activating mutant of the human lutropin receptor. *Mol Cell Endocrinol* 2007;**260–262**:287–93.
44. Zhang M, Tao YX, Ryan GL, Feng X, Fanelli F, Segaloff DL. Intrinsic differences in the response of the human lutropin receptor versus the human follitropin receptor to activating mutations. *J Biol Chem* 2007;**282**:25527–39.
45. Vasseur C, Rodien P, Beau I, Desroches A, Gerard C, de Poncheville L, et al. A chorionic gonadotropin-sensitive mutation in the follicle-stimulating hormone receptor as a cause of familial gestational spontaneous ovarian hyperstimulation syndrome. *N Engl J Med* 2003;**349**:753–9.
46. Smits G, Olatunbosun O, Delbaere A, Pierson R, Vassart G, Costagliola S. Ovarian hyperstimulation syndrome due to a mutation in the follicle-stimulating hormone receptor. *N Engl J Med* 2003;**349**:760–6.
47. Montanelli L, Delbaere A, Di Carlo C, Nappi C, Smits G, Vassart G, et al. A mutation in the follicle-stimulating hormone receptor as a cause of familial spontaneous ovarian hyperstimulation syndrome. *J Clin Endocrinol Metab* 2004;**89**:1255–8.
48. Montanelli L, Van Durme JJ, Smits G, Bonomi M, Rodien P, Devor EJ, et al. Modulation of ligand selectivity associated with activation of the transmembrane region of the human follitropin receptor. *Mol Endocrinol* 2004;**18**:2061–73.
49. De Leener A, Montanelli L, Van Durme J, Chae H, Smits G, Vassart G, et al. Presence and absence of follicle-stimulating hormone receptor mutations provide some insights into spontaneous ovarian hyperstimulation syndrome physiopathology. *J Clin Endocrinol Metab* 2006;**91**:555–62.
50. Angelova K, Fanelli F, Puett D. A model for constitutive lutropin receptor activation based on molecular simulation and engineered mutations in transmembrane helices 6 and 7. *J Biol Chem* 2002;**277**:32202–13.
51. Fanelli F, Verhoef-Post M, Timmerman M, Zeilemaker A, Martens JW, Themmen AP. Insight into mutation-induced activation of the luteinizing hormone receptor: molecular simulations predict the functional behavior of engineered mutants at M398. *Mol Endocrinol* 2004;**18**:1499–508.
52. Zhang M, Mizrachi D, Fanelli F, Segaloff DL. The formation of a salt bridge between helices 3 and 6 is responsible for the constitutive activity and lack of hormone responsiveness of the naturally occurring L457R mutation of the human lutropin receptor. *J Biol Chem* 2005;**280**:26169–76.
53. Fanelli F, De Benedetti PG. Computational modeling approaches to structure–function analysis of G protein-coupled receptors. *Chem Rev* 2005;**105**:3297–351.
54. Feng X, Muller T, Mizrachi D, Fanelli F, Segaloff DL. An intracellular loop (IL2) residue confers different basal constitutive activities to the human lutropin receptor and human thyrotropin receptor through structural communication between IL2 and helix 6, via helix 3. *Endocrinology* 2008;**149**:1705–17.
55. Abell AN, Segaloff DL. Evidence for the direct involvement of transmembrane region 6 of the lutropin/choriogonadotropin receptor in activating Gs. *J Biol Chem* 1997;**272**:14586–91.
56. Tao YX, Johnson NB, Segaloff DL. Constitutive and agonist-dependent self-association of the cell surface human lutropin receptor. *J Biol Chem* 2004;**279**:5904–14.

57. Guan R, Feng X, Wu X, Zhang M, Zhang X, Hebert TE, et al. Bioluminescence resonance energy transfer studies reveal constitutive dimerization of the human lutropin receptor and a lack of correlation between receptor activation and the propensity for dimerization. *J Biol Chem* 2009;**284**:7483–94.
58. Martens JWM, Verhoef-Post M, Abelin N, Ezabella M, Toledo SPA, Brunner HG, et al. A homozygous mutation of the luteinizing hormone receptor causes partial Leydig cell hypoplasia: correlation between receptor activity and phenotype. *Mol Endocrinol* 1998;**12**:775–84.
59. Latronico AC, Chai Y, Arnhold IJP, Liu X, Mendonca BB, Segaloff DL. A homozygous microdeletion in helix 7 of the luteinizing hormone receptor associated with familial testicular and ovarian resistance is due to both decreased cell surface expresssion and impaired effector activation by the cell surface receptor. *Mol Endocrinol* 1998;**12**:442–50.
60. Mizrachi D, Segaloff DL. Intracellularly located misfolded glycoprotein hormone receptors associate with different chaperone proteins than their cognate wild-type receptors. *Mol Endocrinol* 2004;**18**:1768–77.
61. Harwood JP, Conti M, Conn PM, Dufau ML, Catt KJ. Receptor regulation and target cell responses: studies in the ovarian luteal cell. *Mol Cell Endocrinol* 1978;**11**:121–35.
62. Catt KJ, Harwood JP, Clayton RN, Davies TF, Chan V, Katikineni M, et al. Regulation of peptide hormone receptors and gonadal steroidogenesis. *Recent Prog Horm Res* 1980;**36**:557–622.
63. Bruysters M, Christin-Maitre S, Verhoef-Post M, Sultan C, Auger J, Faugeron I, et al. A new LH receptor splice mutation responsible for male hypogonadism with subnormal sperm production in the propositus, and infertility with regular cycles in an affected sister. *Hum Reprod* 2008;**23**:1917–23.
64. Gromoll J, Eiholzer U, Nieschlag E, Simoni M. Male hypogonadism caused by homozygous deletion of exon 10 of the luteinizing hormone (LH) receptor: differential action of human chorionic gonadotropin and LH. *J Clin Endocrinol Metab* 2000;**85**:2281–6.
65. Muller T, Gromoll J, Simoni M. Absence of exon 10 of the human luteinizing hormone (LH) receptor impairs LH, but not human chorionic gonadotropin action. *J Clin Endocrinol Metab* 2003;**88**:2242–9.
66. Richter-Unruh A, Korsch E, Hiort O, Holterhus PM, Themmen AP, Wudy SA. Novel insertion frameshift mutation of the LH receptor gene: problematic clinical distinction of Leydig cell hypoplasia from enzyme defects primarily affecting testosterone biosynthesis. *Eur J Endocrinol* 2005;**152**:255–9.
67. Richter-Unruh A, Martens JW, Verhoef-Post M, Wessels HT, Kors WA, Sinnecker GH, et al. Leydig cell hypoplasia: cases with new mutations, new polymorphisms and cases without mutations in the luteinizing hormone receptor gene. *Clin Endocrinol (Oxf)* 2002;**56**:103–12.
68. Gromoll J, Partsch C-J, Simoni M, Nordhoff V, Sippell WG, Nieschlag E, et al. A mutation in the first transmembrane domain of the lutropin receptor causes male precocious puberty. *J Clin Endocrinol Metab* 1998;**83**:476–80.
69. Evans BAJ, Bowen DJ, Smith PJ, Clayton PE, Gregory JW. A new point mutation in the luteinising hormone receptor gene in familial and sporadic male limited precocious puberty: genotype does not always correlate with phenotype. *J Med Genet* 1996;**33**:143–7.
70. Kraaij R, Post M, Kremer H, Milgrom E, Epping W, Brunner HG, et al. A missense mutation in the second transmembrane segment of the luteinizing hormone receptor causes familiar male precocious puberty. *J Clin Endocrinol Metab* 1995;**80**:3168–72.
71. Yano K, Kohn LD, Saji M, Kataoka N, Okuno A, Cutler Jr GB. A case of male-limited precocious puberty caused by a point mutation in the second transmembrane domain of the luteinizing hormone choriogonadotropin receptor gene. *Biochem Biophys Res Commun* 1996;**220**:1036–42.

72. Ignacak M, Hilczer M, Zarzycki J, Trzeciak WH. Substitution of M398T in the second transmembrane helix of the LH receptor in a patient with familial male-limited precocious puberty. *Endocr J* 2000;**47**:595–9.
73. Kremer H, Martens JW, van Reen M, Verhoef-Post M, Wit JM, Otten BJ, et al. A limited repertoire of mutations of the luteinizing hormone (LH) receptor gene in familial and sporadic patients with male LH-independent precocious puberty. *J Clin Endocrinol Metab* 1999;**84**:1136–40.
74. Latronico AC, Anasti J, Arnhold IJP, Mendonca BB, Domenice S, Albano MC, et al. A novel mutation of the luteinizing hormone receptor gene causing male gonadotropin-independent precocious puberty. *J Clin Endocrinol Metab* 1995;**80**:2490–4.
75. Kremer H, Mariman E, Otten BJ, Moll Jr GW, Stoelinja GB, Wit JM, et al. Cosegregation of missense mutations of the luteinizing hormone receptor gene with familial male-limited precocious puberty. *Hum Mol Genet* 1993;**2**:1779–83.
76. Kosugi S, Van Dop C, Geffner ME, Rabl W, Carel JC, Chaussain JL, et al. Characterization of heterogeneous mutations causing constitutive activation of the luteinizing hormone receptor in familial male precocious puberty. *Hum Mol Genet* 1995;**4**:183–8.
77. Yano K, Saji M, Hidaka A, Moriya N, Okuno A, Kohn LD, et al. A new constitutively activating point mutation in the luteinizing hormone/choriogonadotropin receptor gene in cases of male-limited precocious puberty. *J Clin Endocrinol Metab* 1995;**80**:1162–8.
78. Laue L, Wu SM, Kudo M, Hsueh AJW, Cutler GB, Jelly DH, et al. Heterogeneity of activating mutations of the human luteinizing hormone receptor in male-limited precocious puberty. *Biochem Mol Med* 1996;**58**:192–8.
79. Cocco S, Meloni A, Marini MG, Cao A, Moi P. A missense (T577I) mutation in the luteinizing hormone receptor gene associated with familial male-limited precocious puberty. *Hum Mutat* 1996;**7**:164–6.
80. Yano K, Hidaka A, Saji M, Polymeropoulos MH, Okuno A, Kohn LD, et al. A sporadic case of male-limited precocious puberty has the same constitutively activating point mutation in luteinizing hormone/choriogonadotropin receptor gene as familial cases. *J Clin Endocrinol Metab* 1994;**79**:1818–23.
81. Kawate N, Kletter GB, Wilson BE, Netzloff ML, Menon KMJ. Identification of constitutively activating mutation of the luteinising hormone receptor in a family with male limited gonadotrophin independent precocious puberty (testotoxicosis). *J Med Genet* 1995;**32**:553–4.
82. Muller J, Gondos B, Kosugi S, Mori T, Shenker A. Severe testotoxicosis phenotype associated with Asp578->Tyr mutation of the lutrophin/choriogonadotrophin receptor gene. *J Med Genet* 1998;**35**:340–1.
83. Wu SM, Leschek EW, Brain C, Chan WY. A novel luteinizing hormone receptor mutation in a patient with familial male-limited precocious puberty: effect of the size of a critical amino acid on receptor activity. *Mol Genet Metab* 1999;**66**:68–73.
84. Canto P, Soderlund D, Ramon G, Nishimura E, Mendez JP. Mutational analysis of the luteinizing hormone receptor gene in two individuals with Leydig cell tumors. *Am J Med Genet* 2002;**108**:148–52.
85. Wu SM, Hallermeier KM, Laue L, Brain C, Berry C, Grant DB, et al. Inactivation of the luteinizing hormone/chorionic gonadotropin receptor by an insertional mutation in Leydig cell hypoplasia. *Mol Endocrinol* 1998;**12**:1651–60.
86. Leung MY, Steinbach PJ, Bear D, Baxendale V, Fechner PY, Rennert OM, et al. Biological effect of a novel mutation in the third leucine-rich repeat of human luteinizing hormone receptor. *Mol Endocrinol* 2006;**20**:2493–503.
87. Misrahi M, Meduri G, Pissard S, Bouvattier C, Beau I, Loosfelt H, et al. Comparison of immunocytochemical and molecular features with the phenotype in a case of incomplete male pseudohermaphroditism associated with a mutation of the luteinizing hormone receptor. *J Clin Endocrinol Metab* 1997;**82**:2159–65.

88. Richter-Unruh A, Verhoef-Post M, Malak S, Homoki J, Hauffa BP, Themmen AP. Leydig cell hypoplasia: absent luteinizing hormone receptor cell surface expression caused by a novel homozygous mutation in the extracellular domain. *J Clin Endocrinol Metab* 2004;**89**:5161–7.
89. Gromoll J, Schulz A, Borta H, Gudermann T, Teerds KJ, Greschniok A, et al. Homozygous mutation within the conserved Ala-Phe-Asn-Glu-Thr motif of exon 7 of the LH receptor causes male pseudohermaphroditism. *Eur J Endocrinol* 2002;**147**:597–608.
90. Laue LL, Wu SM, Kudo M, Bourony CJ, Cutler GB, Hsueh AJW, et al. Compound heterozygous mutations of the luteinizing hormone receptor gene in Leydig cell hypoplasia. *Mol Endocrinol* 1996;**10**:987–97.
91. Martens JW, Lumbroso S, Verhoef-Post M, Georget V, Richter-Unruh A, Szarras-Czapnik M, et al. Mutant luteinizing hormone receptors in a compound heterozygous patient with complete Leydig cell hypoplasia: abnormal processing causes signaling deficiency. *J Clin Endocrinol Metab* 2002;**87**:2506–13.
92. Stavrou SS, Zhu Y-S, Cai LQ, Katz MD, Herrara C, DeFillo-Ricart M, et al. A novel mutation of the human luteinizing hormone receptor in 46XY and 46XX sisters. *J Clin Endocrinol Metab* 1998;**83**:2091–8.
93. Pals-Rylaarsdam R, Liu G, Brickman W, Duranteau L, Monroe J, El-Awady MK, et al. A novel double mutation in the luteinizing hormone receptor in a kindred with familial Leydig cell hypoplasia and male pseudohermaphroditism. *Endocr Res* 2005;**31**:307–23.
94. Laue L, Wu SM, Kudo M, Hsueh AJW, Cutler Jr GB, Griffin JE, et al. A nonsense mutation of the human luteinizing hormone receptor gene in Leydig cell hypoplasia. *Hum Mol Genet* 1995;**4**:1429–33.
95. Latronico AC, Anasti J, Arnhold IJP, Rapaport R, Mendonca BB, Bloise W, et al. Testicular and ovarian resistance to luteinizing hormone caused by homozygous inactivating mutations of the luteinizing hormone receptor gene. *N Engl J Med* 1996;**334**:507–12.
96. Toledo SPA, Brunner HG, Kraaij R, Post M, Dahia PLM, Hayashida CY, et al. An inactivating mutation of the luteinizing hormone receptor causes amenorrhea in a 46, XX female. *J Clin Endocrinol Metab* 1996;**81**:3850–4.
97. Kremer H, Kraaij R, Toledo SPA, Post M, Fridman JB, Hayashida CY, et al. Male pseudohermaphroditism due to a homozygous missense mutation of the luteinizing hormone receptor gene. *Nat Genet* 1995;**9**:160–4.

Follicle Stimulating Hormone Receptor Mutations and Reproductive Disorders

Ya-Xiong Tao[*] and
Deborah L. Segaloff[†]

[*]Department of Anatomy, Physiology and Pharmacology, College of Veterinary Medicine, 212 Greene Hall, Auburn University, Auburn, Alabama 36849

[†]Department of Molecular Physiology and Biophysics, The Roy J. and Lucille A. Carver College of Medicine, The University of Iowa, Iowa City, Iowa 52242

I. Introduction	116
II. Follicle Stimulating Hormone Receptor	116
III. Inactivating *FSHR* Mutations and Hypergonadotropic Hypogonadism	118
IV. Gain-of-Function *FSHR* Mutations and Spontaneous Ovarian Hyperstimulation Syndrome	121
V. Structure–Function Insights from Studies of Constitutively Active FSHR Mutants	123
VI. Conclusions	125
References	125

The follicle stimulating hormone receptor (FSHR) plays a critical role in reproductive function. In the males, FSH supports spermatogenesis, whereas in females, FSH is absolutely required for ovarian follicle growth. In females, inactivating mutations in the *FSHR* result in ovarian dysgenesis with amenorrhea and infertility. The few males reported with severe inactivating mutations exhibited varying spermatogenic defects, but not azoospermia. While these findings may potentially suggest that FSH action is not absolutely required for spermatogenesis, it cannot be ruled out that these individuals have some residual FSHR activity. Gain-of-function mutations in the *FSHR* cause spontaneous ovarian hyperstimulation syndrome in females due to the inappropriate stimulation of the mutant FSHR by human choriogonadotropin.

I. Introduction

Follicle stimulating hormone (follitropin, FSH) is one of the glycoprotein hormones produced by the anterior pituitary gland. Under pulsatile stimulation by gonadotropin-releasing hormone, gonadotrophs in the pituitary gland secrete the gonadotropins FSH and luteinizing hormone (LH). FSH, like the other glycoprotein hormones, is a heterodimer consisting of a common α-subunit and a hormone-specific β-subunit.[1] The two subunits are noncovalently associated. The individual subunits are devoid of any gonadotropic activity. When the two subunits are genetically fused together, the fused chimera peptides are biologically active. The addition of carboxy-terminal peptide of human chorionic gonadotropin (hCG) to these chimeras further enhanced their *in vivo* half-life.[2,3] Extensive heterogeneous glycosylation in both subunits results in numerous isoforms.

The crystal structure of a mutant FSH lacking one glycosylation site has been determined.[4] Similar to hCG,[5] the α- and β-subunits consist of central cystine-knot motifs, with the two subunits associated tightly and forming an elongated and slightly curved structure.[4] There are also some differences in the structures of the β-subunits of hCG and FSH that may be important in binding and activation of the cognate receptors.[4]

FSH is essential for gonadal development, maturation and function. In females, the FSHR is expressed on the granulosa cells of growing follicles. As such, the initiation of follicular development is not dependent on FSH.[6] Subsequently, though, follicular growth requires FSH stimulation. In response to FSH, granulosa cells synthesize estradiol, which synergizes with FSH to promote follicular growth and the differentiation of the granulosa cells. In males, the FSHR is expressed on testicular Sertoli cells and the actions of FSH on the FSHR serve to support spermatogenesis.

II. Follicle Stimulating Hormone Receptor

Shortly after the cloning of the TSH receptor (TSHR) and LH/chorionic gonadotropin (CG) receptor (LHCGR),[7,8] the FSHR cDNA was cloned from a rat Sertoli cell cDNA library.[9] Several groups subsequently independently reported the cloning of human (h) FSHR cDNA.[10–12] There were some differences between the reported sequences, including two polymorphisms A307T and S680N.

The predicted hFSHR is a protein of 695 amino acids, with the first 17 amino acids predicted to be the signal peptide. The first 349 amino acids comprise the large extracellular domain (ECD) that is responsible for high-affinity ligand binding.[13,14] The remaining 329 amino acids correspond to the serpentine portion of the receptor, with seven transmembrane domains (TMs) connected by alternating extra- and intracellular loops, ending with the carboxyl tail. The *FSHR* gene consists of 10 exons, with the first nine exons encoding for the large ECD, and exon 10 encoding for the seven TMs and the C-terminal tail. The human *FSHR* gene is localized to chromosome 2p21.[15,16]

The ECD is composed of 10 leucine-rich repeats flanked at either end by a cysteine-rich region. The cysteine-rich region at the carboxyl end of the ECD is also termed as the hinge region because it connects the rest of the ECD to the serpentine domain. A major breakthrough in FSHR research came with the recent report of the crystal structure of hormone-binding region of the hFSHR ECD (FSHR$_{HB}$, representing the ECD without the hinge region) in complex with FSH.[17] The hormone-occupied FSHR$_{HB}$ was observed to form low-affinity dimers both in solution and in crystal form. In the crystal structure, it was determined that one FSH molecule binds to one hFSHR$_{HB}$ monomer in a hand-clasp fashion. The elongated curved hFSHR$_{HB}$ has a large and highly charged interface with both α- and β-subunits of the hormone. Upon binding to the receptor, FSH undergoes a concerted conformational change that results in receptor activation.[17,18] The crystal structures of the ECD containing the hinge region and the full-length FSHR have not yet been reported.

The FSHR is primarily coupled to the stimulatory heterotrimeric G protein Gs. Upon ligand binding, the receptor is stabilized in an active state, which permits activation of Gs and the consequent stimulation of the cAMP/protein kinase A pathway. When recombinant FSHR is expressed at high receptor densities, increased inositol phosphate formation can also be observed with high concentrations of FSH.[19,20] However, given the lower expression levels of endogenous FSHR, activation of this pathway may not normally occur.

As noted earlier, structural studies on the hormone-occupied FSHR$_{HB}$ show the complex to form low-affinity dimers through interactions between the ECDs.[17,18] Studies on the unoccupied full-length FSHR using resonance energy transfer techniques further suggest constitutive dimerization of the full-length receptor that initiates early in the biosynthetic pathway[21] (Guan *et al.*, manuscript submitted). Although activation of the FSHR does not affect its propensity for dimerization, it cannot be ruled out that receptor activation may cause structural rearrangements within the dimeric complex. The obligate and constitutive nature of FSHR dimerization is in agreement with similar findings on the related lutropin and thyrotropin receptors.[22–24]

III. Inactivating *FSHR* Mutations and Hypergonadotropic Hypogonadism

Inactivating FSHR mutations are autosomal recessive and clinical phenotypes are therefore only observed in homozygous and compound heterozygous individuals. Since in females FSH is absolutely required for follicle growth, severe loss-of-function mutations in the *FSHR* in females result in ovarian dysgenesis with amenorrhea and infertility. The first reported inactivating mutation of the *FSHR* is A189V in the ECD.[25] This mutation is associated with primary amenorrhea and streak ovaries with primordial and primary follicles but no follicles of more advanced stages. The mutation is very frequent in the Finnish population[26] but was not found in patients from Germany,[27] England and France,[28] and North America.[29] When inactivating FSHR mutants retain partial function, the phenotype is less severe.[28,30] Beau *et al.* reported *FSHR* mutations associated with secondary amenorrhea and high concentrations of gonadotropins (particularly FSH). In this case, follicular development was arrested at the small antral stage (maximal diameter of 5 mm) and the patient was found to have the compound heterozygous mutations I160T and R573C.[28] The same group subsequently described another patient with compound heterozygous mutations in *FSHR*. This patient with primary amenorrhea also only had small antral follicles (maximal diameter of 3 mm) in her ovaries. She was a compound heterozygote for D224V and L601V.[30] Both of these patients had normal pubertal development, compared with delayed puberty in the majority of patients with A189V *FSHR*.

Since the original report of Aittomaki *et al.*, nine additional inactivating *FSHR* mutations have been reported, including I160T,[28] N191I,[31] G221V,[32] D224V,[30] and P348R[33] (all in the ECD), A419T in TM2,[34] P519T in EL2,[35] R573C in IL3,[28] and L601V in EL3.[30]

In contrast to females, inactivating *FSHR* mutations in males do not necessarily cause infertility. Males homozygous for the *FSHR* mutation A189V exhibited variable degrees of spermatogenesis.[36] They did not show azoospermia or absolute infertility, unlike the female patients who were all infertile. Indeed, some of the patients fathered children. However, FSH was found to be necessary to maintain normal testicular size and quantitatively normal spermatogenesis. The analysis of the male patients harboring FSHR A189V may suggest that, in contrast to females, FSH may not be as essential for pubertal initiation and maintenance of spermatogenesis as previously thought. The other possibility, though, is that a very low residual level of activity of the FSHR in testicular Sertoli cells may be all that is necessary for spermatogenesis. Something to be considered is that the testes, when properly descended in the scrotum (as in the A189V males), are at a reduced temperature compared to the body's core temperature. It has been previously shown that inactivating rat LHCGR mutants that are misfolded and retained in the endoplasmic reticulum (ER) can be partially rescued by decreased temperatures.[37] Therefore, one can speculate that males with a misfolded inactivating FSHR such as A189V may have a small degree of "rescued" cell surface FSHR expression.

In the patients with inactivating *FSHR* mutations, serum gonadotropin levels are increased due to lack of negative feedback. Therefore, this condition is referred to as hypergonadotropic hypogonadism, different from a defect in FSH secretion.[38,39]

In a study trying to identify *FSHR* mutations from males with idiopathic infertility, no functional mutations were identified.[40] One mutation identified, V341A, was without functional effect. The two polymorphic variants, A307T and S680N, also functioned normally. The two haplotypes, A307-S680 and T307-N680, were distributed equally in fertile and infertile males.[40] Therefore, *FSHR* mutation is not common in males with idiopathic infertility.

In the original report of the first inactivating mutation of the *FSHR*, the wild-type (WT) or mutant FSHR was expressed in MSC-1 cells, a mouse Sertoli cell line that does not express endogenous FSHR.[25] These experiments showed that A189V does not respond to FSH stimulation with increased cAMP production. Ligand-binding experiments showed that although the mutant receptor binds with normal affinity, binding capacity was dramatically reduced.[25] Expression of A189V in porcine granulosa cells also showed that it is not functional.[41] Later studies showed that when expressed in COS-7 (a monkey kidney cell line) and mouse granulosa tumor cell line KK-1, the WT FSHR could be readily seen in nonpermeabilized cells, but A189V could not be detected on the cell surface. Permeabilization showed that the mutant receptor is expressed but retained intracellularly.[42] Therefore, according to the classification scheme we proposed,[43–45] A189V is a Class II mutant. Indeed, the corresponding mutation in human LHCGR also results in intracellular retention and consequently diminished signaling in the cell.[42]

Misrahi and colleagues performed detailed functional characterization of several FSHR mutants.[28,30] They showed that when expressed in COS-7 cells, I160T and D224V had negligible binding to iodinated FSH; therefore, no signaling could be measured. Confocal experiments showed that these mutants were retained intracellularly. Immunoblotting experiments showed that D224V is sensitive to endglycosidase H treatment, suggesting that it is composed of immature receptor that had not exited the ER. No mature form of the receptor could be detected. Therefore, both mutants belong to Class II. R573C and L601V bound FSH with similar affinities as WT FSHR; however, they had impaired signaling, with maximal responses of 30% and 12% of the WT FSHR, respectively.[28,30] Hence, these mutants are Class IV mutants that are defective in signaling.[45] The mechanisms underlying the impairments in ligand-induced signaling by these two mutants have not been investigated further. R573 is located in the cytoplasmic end of TM6 and the third intracellular loop, therefore, potentially involved in G protein coupling and activation.[46,47] L601 is located in the third extracellular loop precluding a direct role in G protein coupling and activation. It was speculated that this residue might be important for maintaining a proper conformation for the third extracellular loop and TM6 and TM7.[30]

A correlation between the residual activity of inactivating *FSHR* mutations and the severity of the clinical phenotype has been observed.[30] One female patient with higher FSHR residual activity had larger follicles, better responses to FSH stimulation with increased serum estrogen and inhibin levels and follicle growth (5–8.3 mm). The second patient with minimal residual activity in FSHR did not respond to FSH stimulation with increased serum estrogen and inhibin levels or follicle growth.[30] Consistent differences in basal serum estrogen and inhibin levels were also observed between these two patients with the first patient having higher basal serum levels of these two hormones.

P348R has a complete absence of binding and signaling.[33] However, it is not known whether the mutant receptor is expressed on the cell surface. Therefore, the mutant might be defective in trafficking to the cell surface or impaired in ligand-binding *per se*. P519T is retained intracellularly, resulting in minimal binding and absent signaling.[35] A419T has normal ligand-binding affinity and capacity, but is defective in signaling,[34] and is therefore a Class IV mutant.[45] A419 lies at the extracellular end of TM2. The mechanism for its effect on signaling remains to be elucidated.

The cause for the intracellular retention of most inactivating FSHR mutants is presumably the misfolding of the receptor induced by the mutation. Consequently, the misfolded receptor is recognized as such by the cell's quality control mechanism in the ER and prevented the export from the ER. The degree to which a misfolded mutant is retained in the ER would be reflective of the severity of its abnormal folding. Small molecule ligands that can pass the cell membrane have been shown to have the potential of acting as pharmacological chaperones rescuing mutant receptors retained in the ER. This has been shown for naturally occurring mutations in the V2 vasopressin receptor,[48] rhodopsin,[49,50] gonadotropin-releasing hormone receptor,[51,52] and the melanocortin-4 receptor,[53] as well as WT or laboratory-generated mutants in several other G-protein-coupled receptors (GPCRs).[54–60] These molecules can cross the cell surface and ER membranes, bind to the misfolded receptors in the ER, and act as folding templates allowing the mutants to fold into native conformations that can exit from the ER. Recently, it was shown that indeed pharmacological chaperones have clinical utility.[61]

Conn and colleagues tested this strategy with the FSHR.[62] They choose an Organon compound, Org41841, which activates the LHCGR,[63] but not the FSHR, at sub-millimolar concentrations. This thienopyr(im)idine binds to the transmembrane regions and the second extracellular loop of the LHCGR and TSHR, distinct from the glycoprotein hormones.[64] When this compound was incubated with WT FSHR, binding capacity and maximal signaling were increased dose-dependently.[62] The compound can also rescue A189V, the first naturally occurring FSHR mutant identified, although several other FSHR mutants could not be rescued.[62]

IV. Gain-of-Function *FSHR* Mutations and Spontaneous Ovarian Hyperstimulation Syndrome

Because of the critical importance of FSH in regulating cell proliferation as well as the fact that some activating mutations in the *TSHR* and *LHCGR* cause thyroid tumors and Leydig cell adenoma, several groups had hypothesized that constitutively active mutations in the *FSHR* might similarly cause granulosa cell tumor formation. However, no germline or somatic mutations of the *FSHR* were found in studies examining granulosa cell tumors.[65–68]

Gromoll and colleagues described the first constitutively active mutation in *FSHR* in a hypophysectomized, but fertile, male.[69] This patient, although lacking gonadotropins, had normal testis volume and spermatogenesis when treated with androgen replacement therapy alone and he fathered three children. Because restoration of spermatogenesis in hypogonadal males typically requires treatment with FSH and testosterone,[70] the authors hypothesized that this patient might have an activating mutation in the *FSHR* that allows sufficient signaling through the FSHR in the absence of FSH. Sequencing showed that indeed this patient harbored a heterozygous D567G (6.30) mutation (the number in parenthesis is the numbering according to the Ballesteros and Weinstein system[71]). Functional studies showed that the mutant receptor resulted in a small but consistent 1.5-fold increase in basal cAMP level compared with WT FSHR.[69] When the mutant was expressed in a mouse Sertoli cell line, a threefold increase in basal cAMP level was observed.[27]

Other activating mutations of the FSHR have been identified in the course of studies examining the *FSHR* for mutations associated with spontaneous ovarian hyperstimulation syndrome (sOHSS).[72–76] Normally, there is relatively strict specificity in the interaction of gonadotropins with their cognate receptors. However, mutations in the receptors might result in relaxing of this stringency. An earlier example of this is a *TSHR* mutation (K183R) in the extracellular domain that results in increased response to hCG by the mutant TSHR, resulting in gestational hyperthyroidism.[77] It was hypothesized that, similarly, mutations in the FSHR might permit the promiscuous stimulation of the mutant FSHR by the high concentrations of hCG that occur during pregnancy, thus causing sOHSS. (Most cases of OHSS are iatrogenic, resulting from overstimulation of the ovaries due to the hormonal stimulation associated with *in vitro* fertilization procedures; only a small percentage of OHSS cases are spontaneous.) To date, six heterozygous mutations of the *FSHR* have been identified associated with sOHSS.[72–76,78] As predicted, these mutant FSHRs, when expressed in heterologous cells, are activated by hCG at lower concentrations than the WT FSHR (although the concentrations of hCG required to activate the mutant FSHRs are still much higher than those needed to stimulate the LHCGR).

The first five *FSHR* mutations thus identified (T449A (3.32), T449I, I545T (5.45), D567G (6.30), and D567N) are all located in the serpentine region of the receptor.[72–76,78] Increases in hCG-binding affinity to the mutants could not be observed. Interestingly, these five mutants also exhibited constitutive activity and the ability to be stimulated by high concentrations of TSH. Indeed, it was further shown that there was a strong correlation between FSHR mutants having constitutive activity and their promiscuous activation by hCG and TSH.[75,79] It has been hypothesized that these FSHR mutations most likely decrease the energy barrier required for activation of the FSHR by low-affinity binding of hCG or TSH. Thus, whereas both the WT and constitutively active FSHR may bind hCG or TSH with very low affinities, these interactions result in the stabilization of the serpentine region of the constitutively active FSHR, but not the WT FSHR, in a more active conformation.[80,81] It should be noted that the association of constitutive activity of the FSHR with its promiscuous activation is unique and not shared by constitutively active mutants of the structurally related LHCGR or TSHR.[75,79]

Recently, a S128Y mutation in the ECD of the FSHR has been identified in a patient with sOHSS.[78] In this case, the increased sensitivity of the FSHR mutant toward hCG is attributable to an increase in the binding affinity of the receptor for hCG. Notably, the S128Y mutant is not stimulated by high concentrations of TSH nor is it constitutively active. These findings demonstrate that sOHSS is not caused by constitutive activity of the FSHR, but rather by mutations that increase the sensitivity of the FSHR to hCG. As these different mutations illustrate, the mechanisms leading to increased sensitivity of the FSHR to hCG, however, may vary.

Polymorphisms at the FSHR might also be related to sOHSS. Delbaere and colleagues showed that in a small cohort of patients, the common polymorphism S680N is related to the severity in iatrogenic sOHSS, with the N680 allele associated with a more severe phenotype[82] (because of almost complete linkage disequilibrium, the allele at codon 307 is threonine).

Whereas the stringency of hormone selective activation of the FSHR can be modified by mutations of the FSHR, the binding selectivity can also be breached by abnormally high concentrations of hCG or TSH. Thus, molar pregnancy or choriocarcinoma, both of which are characterized by excessively high concentrations of hCG, can cause theca lutein cysts.[76,83,84] It may also be that some sporadic cases of sOHSS without *FSHR* mutation might be explained by the promiscuous activation of WT FSHR by the extremely high hCG levels during the first trimester of pregnancy.[78] Furthermore, gonadal hyperstimulation that is associated with severe juvenile primary hyperthyroidism may also be caused by stimulation of the WT hFSHR by the excessively high concentrations of TSH that occur in these cases.[85]

V. Structure–Function Insights from Studies of Constitutively Active FSHR Mutants

The FSHR, together with the TSHR and LHCGR, is member of a subfamily of GPCRs that contain a large N-terminal ECD. These glycoprotein hormone receptors, although sharing significant homology, especially in the TM helices, have different basal activities. In the WT receptors, TSHR has the highest basal activity, followed by LHCGR, with the FSHR having the lowest basal activity. Their susceptibilities to mutation-induced constitutive activation follow the same rank order.

Hsueh and colleagues first showed that human FSHR (D567G), unlike the corresponding mutations in the TSHR and LHCGR, was not constitutively active.[86] Several additional mutants in TM6 of the FSHR, including D581G, D581Y, and C584R, were also not constitutively active.[86] These results suggest that the FSHR or TM6 of the FSHR is more resistant to mutation-induced constitutive activation. After we identified a constitutively active L457R mutation in the *LHCGR* from a boy with gonadotropin-independent precocious puberty,[87] we generated the corresponding mutation in the FSHR. Our data showed that this FSHR mutant has robust constitutive activity, increasing basal cAMP level approximately fivefold above that of the WT FSHR.[88]

To systematically investigate the susceptibility of FSHR to mutation-induced constitutive activation, we recently performed a detailed comparison of the susceptibility to mutation-induced constitutive activation in the LHCGR versus the FSHR.[79] Based on data from the LHCGR, nine homologous mutations, including A376V, M401T, V543L, I545L, A571V, M574I, A575V, I578L, and T580I, were made in the FSHR. These experiments showed that only three FSHR mutants, M401T, I545L, and T580I, have significant constitutive activities. Quantitatively comparing the constitutive activities of these FSHR mutants to the corresponding LHCGR mutants demonstrated that the FSHR is intrinsically less susceptible to constitutive activation than the LHCGR.[79] However, computational modeling suggests that the mechanisms through which homologous mutations cause constitutive activation are similar between the FSHR and LHCGR.[79]

Interestingly, the FSHRs from rat and human have different susceptibilities to mutation-induced constitutive activation. For example, although both Hsueh and colleagues and we showed that hFSHR(D581G) (6.44) is not constitutively active, we showed that the corresponding mutation in rat (r) FSHR(D580G) caused very robust constitutive activation.[89] Thus, the basal cAMP of rFSHR(D580G) is increased $\sim$ 10-fold compared with the WT rFSHR. We exploited the fact that the rat and human FSHRs are highly homologous, with 89% homology over the full-length receptors, and 95% homology in the TMs to determine the basis for the differing activities of the homologous mutations. First, we generated chimeras that switched the ECDs. Data from these chimeras showed that the ECDs do not affect the D(6.44)G mutation-induced constitutive activation. The chimera with the ECD of hFSHR and TMs of rFSHR were constitutively active when the D(6.44)G was introduced. The chimera with the ECD of rFSHR and the TMs of hFSHR was not constitutively active when the D(6.44)G mutation was introduced. These results suggest that the molecular determinants dictating the two receptors to D(6.44)G mutation-induced constitutive activation lie in the serpentine region of the receptors. By generating reciprocal mutations in TM5, TM6, and TM7, we showed that two residues, one in TM6 and one in TM7, could account for the dramatic difference in activities. The double mutant M576T/H615Y hFSHR, although not constitutively active, becomes constitutively active when the D(6.44)G mutation was introduced. Similarly, when Y614 in rFSHR was mutated to the corresponding residue in hFSHR, histidine, the D(6.44)G mutant receptor was not constitutively active any more. Homology modeling suggests that differences in hydrophobic interactions between TM6 and TM7 might account for the different susceptibilities of the rat versus human FSHR to D(6.44)G mutation-induced constitutive activation.[89] Hsueh's group showed that different interactions between TM5 and TM6 are responsible for the different susceptibility to D(6.30)G mutation-induced constitutive activation in the hFSHR (where the mutant is not constitutively active) and the LHCGR (where the mutant is constitutively active).[86]

Multiple mutagenesis experiments at one locus can identify the side chain requirements for eliciting mutation-induced constitutive activation.[90] Indeed, multiple mutations of some of the FSHR loci identified that result in constitutive activation have been performed. These experiments, together with molecular modeling, suggested that weakening of interhelical locks between TM6 and TM3 or TM6 and TM7 was responsible for the constitutive activity of D567

(6.30) and T449 (3.32) mutants.[75] In addition, there is a postulated ionic lock between D(6.30) and R(3.50), linking the cytoplasmic ends of TM3 and TM6 together.[79,91–94] Any mutation that weakens or disrupts this lock could result in constitutive activation. The recently reported structure of opsin in an active state showed that indeed this ionic lock is broken.[95] For the FSHR, a negatively charged residue is needed at D(6.30) to maintain the receptor in an inactive conformation. Mutations that introduce a basic residue such as Arg or Lys at that site result in high constitutive activities.[75]

VI. Conclusions

Although rare, mutations of the FSHR have provided unique insights into the role of the FSHR in human reproductive physiology and into the molecular mechanisms underlying the activation of the FSHR. The identification of additional severely inactivating FSHR mutations, however, would be very illuminating in order to better clarify the role of the FSHR in spermatogenesis in males.

Acknowledgments

Studies discussed from the authors' laboratories were supported by NIH grants DK068614 and HD22196.

References

1. Pierce JG, Parsons TF. Glycoprotein hormones: structure and function. *Annu Rev Biochem* 1981;**50**:465–95.
2. Fares FA, Suganuma N, Nishimori K, LaPolt PS, Hsueh AJ, Boime I. Design of a long-acting follitropin agonist by fusing the C-terminal sequence of the chorionic gonadotropin beta subunit to the follitropin beta subunit. *Proc Natl Acad Sci USA* 1992;**89**:4304–8.
3. Sugahara T, Sato A, Kudo M, Ben-Menahem D, Pixley MR, Hsueh AJ, et al. Expression of biologically active fusion genes encoding the common alpha subunit and the follicle-stimulating hormone beta subunit. Role of a linker sequence. *J Biol Chem* 1996;**271**:10445–8.
4. Fox KM, Dias JA, Van Roey P. Three-dimensional structure of human follicle-stimulating hormone. *Mol Endocrinol* 2001;**15**:378–89.
5. Lapthorn AJ, Harris DC, Littlejohn A, Lustbader JW, Canfield RE, Machin KJ, et al. Crystal structure of human chorionic gonadotropin. *Nature* 1994;**369**:455–61.
6. Oktay K, Briggs D, Gosden RG. Ontogeny of follicle-stimulating hormone receptor gene expression in isolated human ovarian follicles. *J Clin Endocrinol Metab* 1997;**82**:3748–51.
7. Parmentier M, Libert F, Maenhaut C, Lefort A, Gerard C, Perret J, et al. Molecular cloning of the thyrotropin receptor. *Science* 1989;**246**:1620–2.

8. McFarland KC, Sprengel R, Phillips HS, Kohler M, Rosemblit N, Nikolics K, et al. Lutropin-choriogonadotropin receptor: an unusual member of the G protein-coupled receptor family. *Science* 1989;**245**:494–9.
9. Sprengel R, Braun T, Nikolics K, Segaloff DL, Seeburg PH. The testicular receptor for follicle stimulating hormone: structure and functional expression of cloned cDNA. *Mol Endocrinol* 1990;**4**:525–30.
10. Minegishi T, Nakamura K, Takakura Y, Ibuki Y, Igarashi M. Cloning and sequencing of human FSH receptor cDNA. *Biochem Biophys Res Commun* 1991;**175**:1125–30.
11. Gromoll J, Gudermann T, Nieschlag E. Molecular cloning of a truncated isoform of the human follicle stimulating hormone receptor. *Biochem Biophys Res Commun* 1992;**188**:1077–83.
12. Tilly JL, Aihara T, Nishimori K, Jia XC, Billig H, Kowalski KI, et al. Expression of recombinant human follicle-stimulating hormone receptor: species-specific ligand binding, signal transduction, and identification of multiple ovarian messenger ribonucleic acid transcripts. *Endocrinology* 1992;**131**:799–806.
13. Braun T, Schofield PR, Sprengel R. Amino-terminal leucine-rich repeats in gonadotropin receptors determine hormone selectivity. *EMBO J* 1991;**10**:1885–90.
14. Davis D, Liu X, Segaloff DL. Identification of the sites of N-linked glycosylation on the follicle-stimulating hormone (FSH) receptor and assessment of their role in FSH receptor function. *Mol Endocrinol* 1995;**9**:159–70.
15. Rousseau-Merck MF, Atger M, Loosfelt H, Milgrom E, Berger R. The chromosomal localization of the human follicle-stimulating hormone receptor gene (FSHR) on 2p21–p16 is similar to that of the luteinizing hormone receptor gene. *Genomics* 1993;**15**:222–4.
16. Gromoll J, Ried T, Holtgreve-Grez H, Nieschlag E, Gudermann T. Localization of the human FSH receptor to chromosome 2 p21 using a genomic probe comprising exon 10. *J Mol Endocrinol* 1994;**12**:265–71.
17. Fan QR, Hendrickson WA. Structure of human follicle-stimulating hormone in complex with its receptor. *Nature* 2005;**433**:269–77.
18. Fan QR, Hendrickson WA. Assembly and structural characterization of an authentic complex between human follicle stimulating hormone and a hormone-binding ectodomain of its receptor. *Mol Cell Endocrinol* 2007;**260–262**:73–82.
19. Hipkin RW, Liu X, Ascoli M. Truncation of the C-terminal tail of the follitropin receptor does not impair the agonist- or phorbol ester-induced receptor phosphorylation and uncoupling. *J Biol Chem* 1995;**270**:26683–9.
20. Donadeu FX, Ascoli M. The differential effects of the gonadotropin receptors on aromatase expression in primary cultures of immature rat granulosa cells are highly dependent on the density of receptors expressed and the activation of the inositol phosphate cascade. *Endocrinology* 2005;**146**:3907–16.
21. Thomas RM, Nechamen CA, Mazurkiewicz JE, Muda M, Palmer S, Dias JA. Follice-stimulating hormone receptor forms oligomers and shows evidence of carboxyl-terminal proteolytic processing. *Endocrinology* 2007;**148**:1987–95.
22. Tao YX, Johnson NB, Segaloff DL. Constitutive and agonist-dependent self-association of the cell surface human lutropin receptor. *J Biol Chem* 2004;**279**:5904–14.
23. Urizar E, Montanelli L, Loy T, Bonomi M, Swillens S, Gales C, et al. Glycoprotein hormone receptors: link between receptor homodimerization and negative cooperativity. *EMBO J* 2005;**24**:1954–64.
24. Guan R, Feng X, Wu X, Zhang M, Zhang X, Hebert TE, et al. Bioluminescence resonance energy transfer studies reveal constitutive dimerization of the human lutropin receptor and a lack of correlation between receptor activation and the propensity for dimerization. *J Biol Chem* 2009;**284**:7483–94.

25. Aittomaki K, Lucena JLD, Pakarinen P, Sistonen P, Tapanainen J, Gromoll J, et al. Mutation in the follicle-stimulating hormone receptor gene causes hereditary hypergonadotropic ovarian failure. *Cell* 1995;**82**:959–68.
26. Aittomaki K, Herva R, Stenman U, Kaisa J, Yiostalo P, Hovatta Chapelle A. Clinical features of primary ovarian failure caused by a point mutation in the follicle stimulating hormone receptor gene. *J Clin Endocrinol Metab* 1996;**81**:3722–6.
27. Simoni M, Gromoll J, Nieschlag E. The follicle-stimulating hormone receptor: biochemistry, molecular biology, physiology, and pathophysiology. *Endocr Rev* 1997;**18**:739–73.
28. Beau I, Touraine P, Meduri G, Gougeon A, Desroches A, Matuchansky C, et al. A novel phenotype related to partial loss of function mutations of the follicle stimulating hormone receptor. *J Clin Invest* 1998;**102**:1352–9.
29. Layman LC, Amde S, Cohen DP, Jin M, Xie J. The Finnish follicle-stimulating hormone receptor gene mutation is rare in North American women with 46, XX ovarian failure. *Fertil Steril* 1998;**69**:300–2.
30. Touraine P, Beau I, Gougeon A, Meduri G, Desroches A, Pichard C, et al. New natural inactivating mutations of the follicle-stimulating hormone receptor: correlations between receptor function and phenotype. *Mol Endocrinol* 1999;**13**:1844–54.
31. Gromoll J, Simoni M, Nordhoff V, Behre HM, De Geyter C, Nieschlag E. Functional and clinical consequences of mutations in the FSH receptor. *Mol Cell Endocrinol* 1996;**125**:177–82.
32. Nakamura Y, Maekawa R, Yamagata Y, Tamura I, Sugino N. A novel mutation in exon8 of the follicle-stimulating hormone receptor in a woman with primary amenorrhea. *Gynecol Endocrinol* 2008;**24**:708–12.
33. Allen LA, Achermann JC, Pakarinen P, Kotlar TJ, Huhtaniemi IT, Jameson JL, et al. A novel loss of function mutation in exon 10 of the FSH receptor gene causing hypergonadotrophic hypogonadism: clinical and molecular characteristics. *Hum Reprod* 2003;**18**:251–6.
34. Doherty E, Pakarinen P, Tiitinen A, Kiilavuori A, Huhtaniemi I, Forrest S, et al. A novel mutation in the FSH receptor inhibiting signal transduction and causing primary ovarian failure. *J Clin Endocrinol Metab* 2002;**87**:1151–5.
35. Meduri G, Touraine P, Beau I, Lahuna O, Desroches A, Vacher-Lavenu MC, et al. Delayed puberty and primary amenorrhea associated with a novel mutation of the human follicle stimulating hormone receptor: clinical, histological, and molecular studies. *J Clin Endocrinol Metab* 2003;**88**:3491–8.
36. Tapanainen JS, Aittomaki K, Min J, Vaskivuo T, Huhtaniemi IT. Men homozygous for an inactivating mutation of the follicle-stimulating hormone (FSH) receptor gene present variable suppression of spermatogenesis and fertility. *Nat Genet* 1997;**15**:205–6.
37. Jaquette J, Segaloff DL. Temperature sensitivity of some mutants of the lutropin/choriogonadotropin receptor. *Endocrinology* 1997;**138**:85–91.
38. Matthews CH, Borgato S, Beck-Peccoz P, Adams M, Tone Y, Gambino G, et al. Primary amenorrhoea and infertility due to a mutation in the beta-subunit of follicle-stimulating hormone. *Nat Genet* 1993;**5**:83–6.
39. Layman LC, Lee EJ, Peak DB, Namnoum AB, Vu KV, van Lingen BL, et al. Delayed puberty and hypogonadism caused by mutations in the follicle-stimulating hormone beta-subunit gene. *N Engl J Med* 1997;**337**:607–11.
40. Simoni M, Gromoll J, Hoppner W, Kamischke A, Krafft T, Stahle D, et al. Mutational analysis of the follicle-stimulating hormone (FSH) receptor in normal and infertile men: identification and characterization of two discrete FSH receptor isoforms. *J Clin Endocrinol Metab* 1999;**84**:751–5.
41. Ghadami M, Salama SA, Khatoon N, Chilvers R, Nagamani M, Chedrese PJ, et al. Toward gene therapy of primary ovarian failure: adenovirus expressing human FSH receptor corrects the Finnish C566T mutation. *Mol Hum Reprod* 2008;**14**:9–15.

42. Rannikko A, Pakarinen P, Manna PR, Beau I, Misrahi M, Aittomaki K, et al. Functional characterization of the human FSH receptor with an inactivating Ala189Val mutation. *Mol Hum Reprod* 2002;**8**:311–7.
43. Tao YX, Segaloff DL. Functional characterization of melanocortin-4 receptor mutations associated with childhood obesity. *Endocrinology* 2003;**144**:4544–51.
44. Tao YX. Molecular mechanisms of the neural melanocortin receptor dysfunction in severe early onset obesity. *Mol Cell Endocrinol* 2005;**239**:1–14.
45. Tao YX. Inactivating mutations of G protein-coupled receptors and diseases: structure–function insights and therapeutic implications. *Pharmacol Ther* 2006;**111**:949–73.
46. Abell AN, Segaloff DL. Evidence for the direct involvement of transmembrane region six of the lutropin/choriogonadotropin receptor in activating Gs. *J Biol Chem* 1997;**272**:14586–91.
47. Scheerer P, Park JH, Hildebrand PW, Kim YJ, Krauss N, Choe HW, et al. Crystal structure of opsin in its G-protein-interacting conformation. *Nature* 2008;**455**:497–502.
48. Morello JP, Salahpour A, Laperriere A, Bernier V, Arthus MF, Lonergan M, et al. Pharmacological chaperones rescue cell-surface expression and function of misfolded V2 vasopressin receptor mutants. *J Clin Invest* 2000;**105**:887–95.
49. Noorwez SM, Kuksa V, Imanishi Y, Zhu L, Filipek S, Palczewski K, et al. Pharmacological chaperone-mediated *in vivo* folding and stabilization of the P23H-opsin mutant associated with autosomal dominant retinitis pigmentosa. *J Biol Chem* 2003;**278**:14442–50.
50. Noorwez SM, Malhotra R, McDowell JH, Smith KA, Krebs MP, Kaushal S. Retinoids assist the cellular folding of the autosomal dominant retinitis pigmentosa opsin mutant P23H. *J Biol Chem* 2004;**279**:16278–84.
51. Janovick JA, Maya-Nunez G, Conn PM. Rescue of hypogonadotropic hypogonadism-causing and manufactured GnRH receptor mutants by a specific protein-folding template: misrouted proteins as a novel disease etiology and therapeutic target. *J Clin Endocrinol Metab* 2002;**87**:3255–62.
52. Janovick JA, Goulet M, Bush E, Greer J, Wettlaufer DG, Conn PM. Structure-activity relations of successful pharmacologic chaperones for rescue of naturally occurring and manufactured mutants of the gonadotropin-releasing hormone receptor. *J Pharmacol Exp Ther* 2003;**305**:608–14.
53. Fan ZC, Tao YX. Functional characterization and pharmacological rescue of melanocortin-4 receptor mutations identified from obese patients. *J Cell Mol Med* 2009;10.1111/j.1582-4934.2009.00726.x [in press].
54. Chaipatikul V, Erickson-Herbrandson LJ, Loh HH, Law PY. Rescuing the traffic-deficient mutants of rat μ-opioid receptors with hydrophobic ligands. *Mol Pharmacol* 2003;**64**:32–41.
55. Petaja-Repo UE, Hogue M, Bhalla S, Laperriere A, Morello JP, Bouvier M. Ligands act as pharmacological chaperones and increase the efficiency of δ opioid receptor maturation. *EMBO J* 2002;**21**:1628–37.
56. Fan J, Perry SJ, Gao Y, Schwarz DA, Maki RA. A point mutation in the human melanin concentrating hormone receptor 1 reveals an important domain for cellular trafficking. *Mol Endocrinol* 2005;**19**:2579–90.
57. Van Craenenbroeck K, Clark SD, Cox MJ, Oak JN, Liu F, Van Tol HH. Folding efficiency is rate-limiting in dopamine D4 receptor biogenesis. *J Biol Chem* 2005;**280**:19350–7.
58. Robert J, Auzan C, Ventura MA, Clauser E. Mechanisms of cell-surface rerouting of an endoplasmic reticulum-retained mutant of the vasopressin V1b/V3 receptor by a pharmacological chaperone. *J Biol Chem* 2005;**280**:42198–206.
59. Hawtin SR. Pharmacological chaperone activity of SR49059 to functionally recover misfolded mutations of the vasopressin V1a receptor. *J Biol Chem* 2006;**281**:14604–14.
60. Fortin JP, Dziadulewicz EK, Gera L, Marceau F. A nonpeptide antagonist reveals a highly glycosylated state of the rabbit kinin B1 receptor. *Mol Pharmacol* 2006;**69**:1146–57.

61. Bernier V, Morello JP, Zarruk A, Debrand N, Salahpour A, Lonergan M, et al. Pharmacologic chaperones as a potential treatment for X-linked nephrogenic diabetes insipidus. *J Am Soc Nephrol* 2006;**17**:232–43.
62. Janovick JA, Maya-Nunez G, Ulloa-Aguirre A, Huhtaniemi IT, Dias JA, Verbost P, et al. Increased plasma membrane expression of human follicle-stimulating hormone receptor by a small molecule thienopyr(im)idine. *Mol Cell Endocrinol* 2009;**298**:84–8.
63. van Straten NC, Schoonus-Gerritsma GG, van Someren RG, Draaijer J, Adang AE, Timmers C, et al. The first orally active low molecular weight agonists for the LH receptor: thienopyr(im)idines with therapeutic potential for ovulation induction. *ChemBioChem* 2002;**3**:1023–6.
64. Jaschke H, Neumann S, Moore S, Thomas CJ, Colson AO, Costanzi S, et al. A low molecular weight agonist signals by binding to the transmembrane domain of thyroid-stimulating hormone receptor (TSHR) and luteinizing hormone/chorionic gonadotropin receptor (LHCGR). *J Biol Chem* 2006;**281**:9841–4.
65. Fuller PJ, Verity K, Shen Y, Mamers P, Jobling T, Burger HG. No evidence of a role for mutations or polymorphisms of the follicle-stimulating hormone receptor in ovarian granulosa cell tumors. *J Clin Endocrinol Metab* 1998;**83**:274–9.
66. Ligtenberg MJ, Siers M, Themmen AP, Hanselaar TG, Willemsen W, Brunner HG. Analysis of mutations in genes of the follicle-stimulating hormone receptor signaling pathway in ovarian granulosa cell tumors. *J Clin Endocrinol Metab* 1999;**84**:2233–4.
67. Latronico A, Segaloff D. Naturally occurring mutations of the luteinizing hormone receptor: lessons learned about reproductive physiology and G protein-coupled receptors. *Am J Hum Genet* 1999;**65**:949–58.
68. Giacaglia LR, Kohek MBDF, Carvalho FM, Fragoso MC, Mendonca B, Latronico AC. No evidence of somatic activating mutations on gonadotropin receptor genes in sex cord stromal tumors. *Fertil Steril* 2000;**74**:992–5.
69. Gromoll J, Simoni M, Nieschlag E. An activating mutation of the follicle-stimulating hormone receptor autonomously sustains spermatogenesis in a hyophysectomized man. *J Clin Endocrinol Metab* 1996;**81**:1367–70.
70. Schaison G, Young J, Pholsena M, Nahoul K, Couzinet B. Failure of combined follicle-stimulating hormone-testosterone administration to initiate and/or maintain spermatogenesis in men with hypogonadotropic hypogonadism. *J Clin Endocrinol Metab* 1993;**77**:1545–9.
71. Ballesteros JA, Weinstein H. Integrated methods for the construction of three-dimensional models and computational probing of structure–function relations in G protein-coupled receptors. *Methods Neurosci* 1995;366–428.
72. Vasseur C, Rodien P, Beau I, Desroches A, Gerard C, de Poncheville L, et al. A chorionic gonadotropin-sensitive mutation in the follicle-stimulating hormone receptor as a cause of familial gestational spontaneous ovarian hyperstimulation syndrome. *N Engl J Med* 2003;**349**:753–9.
73. Smits G, Olatunbosun O, Delbaere A, Pierson R, Vassart G, Costagliola S. Ovarian hyperstimulation syndrome due to a mutation in the follicle-stimulating hormone receptor. *N Engl J Med* 2003;**349**:760–6.
74. Montanelli L, Delbaere A, Di Carlo C, Nappi C, Smits G, Vassart G, et al. A mutation in the follicle-stimulating hormone receptor as a cause of familial spontaneous ovarian hyperstimulation syndrome. *J Clin Endocrinol Metab* 2004;**89**:1255–8.
75. Montanelli L, Van Durme JJ, Smits G, Bonomi M, Rodien P, Devor EJ, et al. Modulation of ligand selectivity associated with activation of the transmembrane region of the human follitropin receptor. *Mol Endocrinol* 2004;**18**:2061–73.
76. De Leener A, Montanelli L, Van Durme J, Chae H, Smits G, Vassart G, et al. Presence and absence of follicle-stimulating hormone receptor mutations provide some insights into

spontaneous ovarian hyperstimulation syndrome physiopathology. *J Clin Endocrinol Metab* 2006;**91**:555–62.
77. Rodien P, Bremont C, Sanson ML, Parma J, Van Sande J, Costagliola S, et al. Familial gestational hyperthyroidism caused by a mutant thyrotropin receptor hypersensitive to human chorionic gonadotropin. *N Engl J Med* 1998;**339**:1823–6.
78. De Leener A, Caltabiano G, Erkan S, Idil M, Vassart G, Pardo L, et al. Identification of the first germline mutation in the extracellular domain of the follitropin receptor responsible for spontaneous ovarian hyperstimulation syndrome. *Hum Mutat* 2008;**29**:91–8.
79. Zhang M, Tao YX, Ryan GL, Feng X, Fanelli F, Segaloff DL. Intrinsic differences in the response of the human lutropin receptor versus the human follitropin receptor to activating mutations. *J Biol Chem* 2007;**282**:25527–39.
80. Vassart G, Pardo L, Costagliola S. A molecular dissection of the glycoprotein hormone receptors. *Trends Biochem Sci* 2004;**29**:119–26.
81. Caltabiano G, Campillo M, De Leener A, Smits G, Vassart G, Costagliola S, et al. The specificity of binding of glycoprotein hormones to their receptors. *Cell Mol Life Sci* 2008;**65**:2484–92.
82. Daelemans C, Smits G, de Maertelaer V, Costagliola S, Englert Y, Vassart G, et al. Prediction of severity of symptoms in iatrogenic ovarian hyperstimulation syndrome by follicle-stimulating hormone receptor Ser680Asn polymorphism. *J Clin Endocrinol Metab* 2004;**89**:6310–5.
83. Hershman JM. Human chorionic gonadotropin and the thyroid: hyperemesis gravidarum and trophoblastic tumors. *Thyroid* 1999;**9**:653–7.
84. Check JH, Choe JK, Nazari A. Hyperreactio luteinalis despite the absence of a corpus luteum and suppressed serum follicle stimulating concentrations in a triplet pregnancy. *Hum Reprod* 2000;**15**:1043–5.
85. Ryan GL, Feng X, d'Alva CB, Zhang M, Van Voorhis BJ, Pinto EM, et al. Evaluating the roles of follicle-stimulating hormone receptor polymorphisms in gonadal hyperstimulation associated with severe juvenile primary hypothyroidism. *J Clin Endocrinol Metab* 2007;**92**:2312–7.
86. Kudo M, Osuga Y, Kobilka BK, Hsueh AJW. Transmembrane regions V and VI of the human luteinizing hormone receptor are required for constitutive activation by a mutation in the third intracellular loop. *J Biol Chem* 1996;**271**:22470–8.
87. Latronico AC, Abell AN, Arnhold IJP, Liu X, Lins TSS, Brito VN, et al. A unique constitutively activating mutation in the third transmembrane helix of the luteinizing hormone receptor causes sporadic male gonadotropin independent precocious puberty. *J Clin Endocrinol Metab* 1998;**83**:2435–40.
88. Tao YX, Abell AN, Liu X, Nakamura K, Segaloff DL. Constitutive activation of G protein-coupled receptors as a result of selective substitution of a conserved leucine residue in transmembrane helix III. *Mol Endocrinol* 2000;**14**:1272–82.
89. Tao YX, Mizrachi D, Segaloff DL. Chimeras of the rat and human FSH receptors (FSHRs) identify residues that permit or suppress transmembrane 6 mutation-induced constitutive activation of the FSHR via rearrangements of hydrophobic interactions between helices 6 and 7. *Mol Endocrinol* 2002;**16**:1881–92.
90. Tao YX. Constitutive activation of G protein-coupled receptors and diseases: insights into mechanism of activation and therapeutics. *Pharmacol Ther* 2008;**120**:129–48.
91. Gether U. Uncovering molecular mechanisms involved in activation of G protein-coupled receptors. *Endocr Rev* 2000;**21**:90–113.
92. Angelova K, Fanelli F, Puett D. A model for constitutive lutropin receptor activation based on molecular simulation and engineered mutations in transmembrane helices 6 and 7. *J Biol Chem* 2002;**277**:32202–13.

93. Ballesteros JA, Jensen AD, Liapakis G, Rasmussen SG, Shi L, Gether U, et al. Activation of the β_2-adrenergic receptor involves disruption of an ionic lock between the cytoplasmic ends of transmembrane segments 3 and 6. *J Biol Chem* 2001;**276**:29171–7.
94. Greasley PJ, Fanelli F, Rossier O, Abuin L, Cotecchia S. Mutagenesis and modelling of the α_{1b}-adrenergic receptor highlight the role of the helix 3/helix 6 interface in receptor activation. *Mol Pharmacol* 2002;**61**:1025–32.
95. Park JH, Scheerer P, Hofmann KP, Choe HW, Ernst OP. Crystal structure of the ligand-free G-protein-coupled receptor opsin. *Nature* 2008;**454**:183–7.

The Human Prostacyclin Receptor: From Structure Function to Disease

KATHLEEN A. MARTIN,[*,†]
SCOTT GLEIM,[*]
LARKIN ELDERON,[*]
KRISTINA FETALVERO,[*,†]
AND JOHN HWA[*,‡]

[*]*Department of Pharmacology and Toxicology, Dartmouth Medical School, Hanover, New Hampshire 03755*

[†]*Department of Surgery, Section of Vascular Surgery, Dartmouth Hitchcock Medical Center, Lebanon, New Hampshire 03756*

[‡]*Department of Medicine, Section of Cardiology, Dartmouth Hitchcock Medical Center, Lebanon, New Hampshire 03756*

I. History	134
II. Molecular and Structural Biology	138
A. General Background	138
B. Binding Pocket	139
C. Critical Activation Components	140
D. Cysteines and Palmitoylation–Isoprenylation	142
E. Serines and Receptor Regulation	142
F. Model of Receptor-Bound Prostacyclin (PGI_2) Ligand	143
G. NMR-Based Structures of the hIP Receptor	144
III. Pathophysiology	145
A. Atherothrombosis	145
B. Role of hIP in Inflammation and Pain	150
C. Role of hIP in Parturition	151
IV. Therapeutics	152
A. Selective COX-2 Inhibition (Coxibs)	152
B. COX-1/2 Inhibitors (NSAIDs)	153
C. Pulmonary Hypertension	154
V. Genetic Variants	154
References	157

Thirty years have passed since Vane and colleagues first described a substance, prostanoid X, from microsomal fractions (later called prostacyclin) that relaxed rather than contracted mesenteric arteries. The critical role of prostacyclin in many pathophysiological conditions, such as atherothrombosis, has only recently become appreciated (through receptor knockout mice studies, selective cyclooxygenase-2 inhibition clinical trials, and the discovery of dysfunctional prostacyclin receptor genetic variants). Additionally, important roles in such diverse areas as pain and inflammation, and parturition are being uncovered. Prostacyclin-based therapies, currently used for pulmonary hypertension, are accordingly emerging as possible treatments for such diseases, fueling interests in structure function studies for the receptor and signal transduction pathways in native cells. The coming decade is likely to yield many further exciting advances.

I. History

In 1970s, Vane and his colleagues noted that a prostaglandin precursor substance (initially named prostaglandin X, PGX) from pig aorta, relaxed rather than contracted coeliac and mesenteric arteries.[1-4] This substance was identified using a unique bioassay technique developed by Vane—*the Vane Bioassay Cascade.*[5] PGX appeared to oppose the constrictive actions of thromboxane A_2 (TXA_2) in addition to inhibiting platelet clumping. PGX was very unstable with a half-life of several minutes at 37 °C, being completely destroyed at 100 °C within 15 s.[4] The structure of PGX was later established to be 9-deoxy-6,6alpha-epoxy-delta5-PGF1alpha and was renamed prostacyclin.[6] These elegant groundbreaking studies in addition to others led to the awarding of the Nobel Prize in 1982 to John Vane.

The following summarizes the progress of prostacyclin from discovery to today. Some of these are highlighted on the timeline (Fig. 1).

1976—Arterial wall generates a substance (prostaglandin X) which relaxes mesenteric and celiac arteries and inhibits platelet aggregation.[3]

1976—The chemical structure of prostaglandin X was determined and the use of the word prostacyclin proposed.[6]

1976—A balance between antiaggregatory (prostacyclin) and proaggregatory (thromboxane) substances contributed to maintaining the integrity of vascular endothelium and thrombus formation.[4]

1977—Prostacyclin is an endogenous metabolite of arachidonic acid via PGH2 and is responsible for relaxation of coronary arteries.[7]

1977—Demonstration that both arteries and veins released prostacyclin which is a potent inhibitor of platelet aggregation.[8]

1977—Prostacyclin is produced differentially by arterial wall with most at intimal and least toward adventitia.[9]

1977—Prostacyclin stimulate platelet cAMP production.[10]

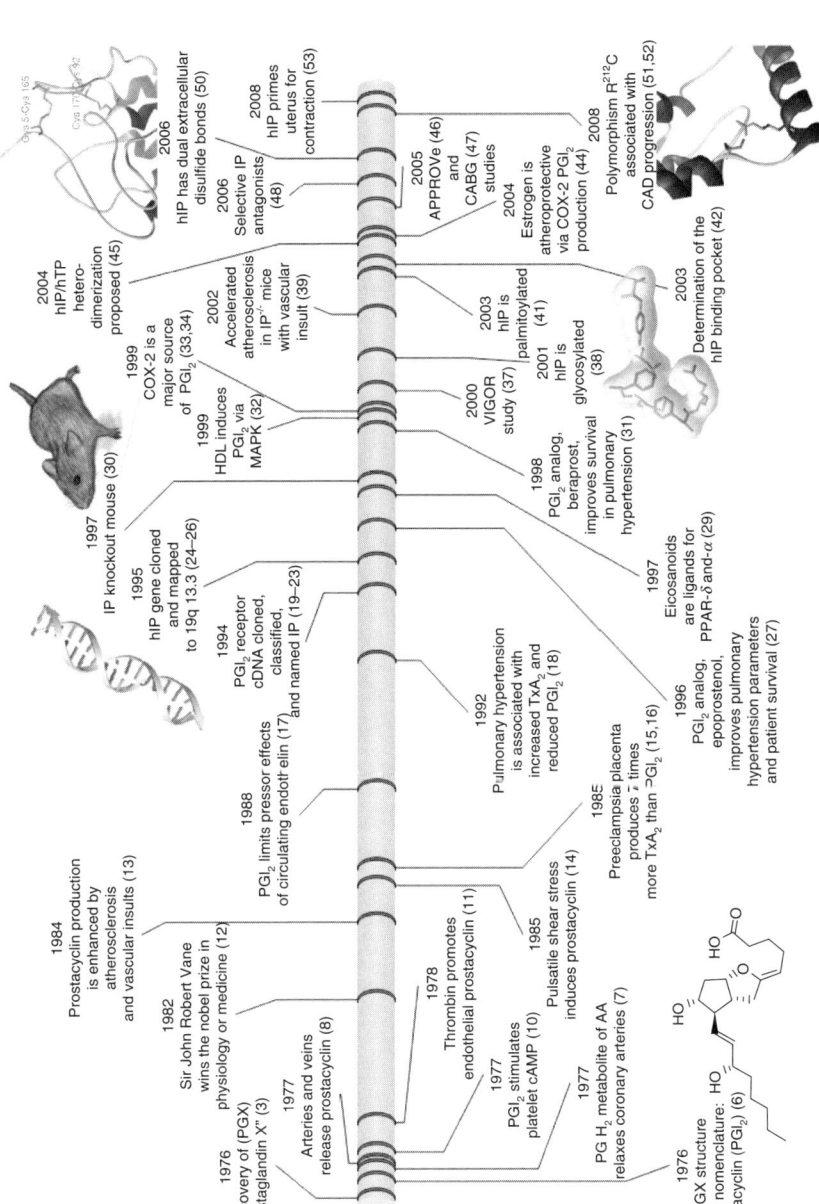

FIG. 1. (Continued)

1978—Thrombin stimulate endothelial production of prostacyclin.[11]
1982—John Vane wins Nobel Prize for discovery of prostacyclin.[12]
1984—Prostacyclin production is enhanced in patients with severe atherosclerosis as a consequence of platelet interactions with endothelium or other vascular insults.[13]
1985—Pulsatile shear stress-induced production of prostacyclin from endothelial cells.[14]
1985—During preeclampsia the placenta produces seven times more thromboxane than prostacyclin.[15,16]
1988—The pressor effects of circulating endothelin are limited in part by the release of prostacyclin.[17]
1992—Thromboxane release is increased and prostacyclin reduced in patients with primary and secondary forms of pulmonary hypertension. This may play a role in the development and maintenance of the disorder.[18]
1994—Cloning and expression of the cDNA for the human prostacyclin receptor[19-21] and mouse prostacyclin receptor.[22]
1994—IUPHAR classifies prostanoid receptors and proposes naming it IP.[23]
1995—Cloning and characterization of the human prostacyclin receptor.[24]
1995—Mapping of chromosomal localization of PTGIR (prostacyclin receptor gene) to chromosome 19 (19q13.3)[25] and genomic organization.[26]
1996—Addition of epoprostenol (IV infusion) improved symptoms, hemodynamic parameters and survival in patients with pulmonary hypertension compared to conventional therapy alone.[27]
1996—The human prostacyclin receptor homologous desensitization is mediated by agonist induce phosphorylation by protein kinase C (PKC).[28]
1997—Eicosanoids are ligands for peroxisome proliferators-activated receptors (PPARs) alpha and delta.[29]
1997—First knockout IP mouse showed increased susceptibility to thrombosis and reduced inflammatory and pain response.[30]
1999—Oral beraprost may improve survival in patients with pulmonary hypertension.[31]
1999—High-density lipoprotein (HDL) induces prostacyclin release via a mitogen-activated protein kinase (MAPK)-dependent pathway.[32]

FIG. 1. Timeline describing a brief history of some of the major landmarks in the study of prostacyclin from its discovery by Vane and his colleagues to recent interesting insights into roles played in normal physiology and pathology. It can be seen that the initial series of foundation studies were followed by a period of relative quiescence until the gene (PTGIR) was cloned, knockout mice studies performed, and COX-2 inhibition appreciated. The increasing interest in genetic variants will likely lead to many new insights into the role played by the prostacyclin receptor in human pathophysiology. (See Color Insert.)

1999—Cyclooxygenase-2 (COX-2) is a major source of systemic prostacyclin biosynthesis in healthy human subjects.[33,34]

2000—Long-term treatment of aerosolized iloprost is effective (hemodynamics and exercise capacity) in treatment of pulmonary hypertension.[35]

2000—Both COX-1 and COX-2 contribute to the formation of prostacyclin in patients with atherosclerosis.[36]

2000—Results of the VIGOR (Vioxx Gastrointestinal Outcome Research) study group showing rofecoxib (selective COX-2 inhibitor) in rheumatoid arthritis patients reduced gastrointestinal (GI) side effects compared to naproxen but showed a higher incidence of myocardial infarction.[37]

2001—The human prostacyclin receptor is glycosylated.[38]

2002—Prostacyclin modulates and limits the response of thromboxane as observed with receptor knockout mice studies. The cardiovascular events associated with selective COX-2 inhibitors may be associated with inhibition of prostacyclin and not thromboxane.[39]

2002—Description and characterization of the first human prostacyclin receptor polymorphism.[40]

2003—The human prostacyclin receptor is palmitoylated.[41]

2003—Determination of the ligand-binding pocket for the human prostacyclin receptor.[42]

2003—NMR techniques used to study ligand recognition sites on the prostacyclin receptor.[43]

2004—Estrogen upregulates prostacyclin via COX-2 leading to atheroprotection.[44] Selective COX-2 inhibitors could remove that protection in premenopausal women.

2004—Heterodimerization of the human prostacyclin with the thromboxane receptor proposed as a means of cross-regulation.[45]

2005—Increased cardiovascular risk associated with the use of rofecoxib in a colorectal adenoma chemoprevention trial.[46]

2005—The use of parecoxib and valdecoxib (two selective COX-2 inhibitors) was associated with increased incidence of cardiovascular events after cardiac surgery.[47]

2006—Development and characterization of selective antagonists RO1138452 and RO3244794.[48]

2006—Prostacyclin receptor induces human vascular smooth muscle cell differentiation via protein kinase A (PKA) pathway.[49]

2006—The human prostacyclin receptor has dual disulfide bonds.[50]

2008—Prostacyclin receptor polymorphism R212C is associated with increased coronary disease and cardiovascular events.[51,52]

2008—Prostacyclin induces uterine contractile proteins in preparation for labor.[53]

In the ensuing years, the receptor was cloned and characterized (Fig. 1). Surprisingly, despite the potential importance in the cardiovascular system, prostacyclin and its analogs have thus far led only to therapy for pulmonary hypertension.[54–57] Further investigation into therapeutic roles for prostacyclin in the cardiovascular disease (CVD), particularly of the coronary arteries was limited by the generally held belief that any deficiencies in prostacyclin pathways would be well compensated for by the nitric oxide (NO) and other vasodilatory systems. However, initial evidence from knockout mice studies,[39,44] followed by human COX-2 inhibition studies,[37,46,58] and, most recently, studies on genetic variants of the human prostacyclin receptor,[51,52] have proved that prostacyclin plays a critical uncompensated role that is particularly evident in the presence of cardiovascular risk factors.[51,59]

II. Molecular and Structural Biology

A. General Background

The human prostacyclin receptor gene (PTGIR) is composed of approximately 7000 bases on chromosome 19 (locus 19q13.3) and contains three exons (two coding) separated by two introns.[26] The PTGIR gene product (human prostacyclin receptor, hIP; International Union of Pharmacology nomenclature) is a seven-transmembrane (TM) G-protein-coupled receptor (GPCR—Class A GPCR). The 386 amino acids encode a short N-terminal tail, three extracellular loops, seven TM-spanning alpha helices, three intracellular loops, with a fourth loop resulting from palmitoylation of intracellular cysteines (308 and 311), and a long intracellular C-terminal tail (Fig. 2). It ranges in molecular mass from 37 to 41 kDa depending on different glycosylation states.[19] Two NXS/T (Asn-X-Ser/Thr) consensus sequences are glycosylated (at positions N7 and N78), facilitating trafficking, ligand binding, and activation.[38] Dual disulfide bonds, one of which is conserved among GPCRs (C92–C170) and another which is a nonconserved bond (C5–C165), stabilize the receptor, enabling proper expression, binding, and activation.[60,61]

Prostacyclin activation of hIP activates membrane-bound adenylyl cyclase (AC) and subsequent formation of the second messenger cyclic adenosine monophosphate (cAMP)[62,63] (Fig. 3). This signaling cascade triggers a host of cellular responses including the classically described inhibition of platelet aggregation and promotion of vascular smooth muscle cell (VSMC) relaxation.[4] The prostacyclin receptor (IP) has also been reported to couple to Gq to mobilize calcium in heterologous expression systems,[64] but it is not yet known whether this occurs in native human VSMCs. At high concentrations, prostacyclin agonists can activate the nuclear receptors PPAR-α and -δ.[29,65,66]

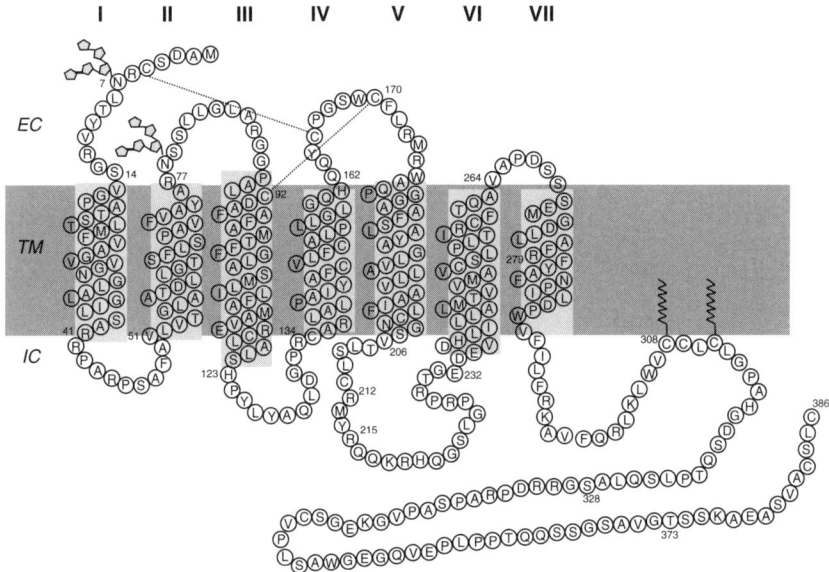

FIG. 2. The secondary structure of the human prostacyclin receptor showing the seven TM domains with an extracellular (EC) N terminus and an intracellular (IC) C terminus. The extracellular domain is characterized by site of glycosylation (chains of pentagons) and dual disulfide bonds (dashed lines). Highly conserved residues critical for ligand binding are located in the transmembrane domain, along with critical residues necessary for the structural transmission of the activating conformational changes. The C-terminal tail is highlighted by dual palmitoylation (serrated lines) and multiple sites of phosphorylation.

Implications for such signaling in the vasculature are only beginning to be appreciated, but a role for prostacyclin PPAR signaling in promoting angiogenesis has been suggested based on prostacyclin analogs selective for nuclear versus cell surface receptors.[67]

B. Binding Pocket

The putative prostacyclin receptor ligand-binding pocket is located in the upper half of the TM-spanning region[42,68] (Fig. 4). Earlier studies of chimeric mouse prostaglandin D and I receptors (mDP and mIP) suggest the involvement of TM domain VI and TMVII in specific substrate recognition, while TMI and portions of the first extracellular loop are required for broader substrate recognition.[69,70] Recent NMR studies have also suggested the presence of a ligand recognition pocket within the extracellular domain.[43] Using both mutagenesis and molecular modeling, Stitham et al. identified four amino acid residues in the putative binding pocket critical for receptor–ligand recognition. The highly conserved R279 (TMVII) (100% conservation across all prostanoid

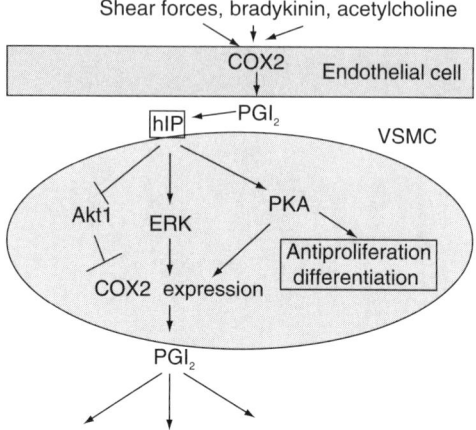

FIG. 3. Prostacyclin signal transduction leading to vascular smooth muscle cell (VSMC) differentiation and reduced proliferation. Additionally, through the PKA, ERK, and AkT1 pathways there is induction of further prostacyclin release leading to antiproliferation and differentiation (quiescent phenotype) on adjacent layers of VSMC. The signaling is likely to be far more complex with the development of further insights into the multiple roles played by prostacyclin in human pathophysiology.

receptors) serves as a positive counter ion to the C1-carboxylate group of PGI_2. The adjacent F278 (TMVII) participates in hydrophobic interactions with the central oxalane ring and the α-tail of the ligand. Y75 (TMII) forms hydrogen bonds with the C11-hydroxyl group of PGI_2, and F95 (TMIII) acts as a nonpolar sidewall limiting free rotation of the ω-tail of PGI_2.[42]

C. Critical Activation Components

Residues within the TM domain provide the conformational switch by which the ligand-binding-induced signal is transduced through the receptor protein to cytoplasmic effector molecules. Using site-directed mutagenesis of conserved and nonconserved residues in the TM domain, Stitham et al. further identified four distinct clusters critical for these activating conformational changes: (1) an immediate binding-pocket cluster, including ligand-interacting residues R279, F95, and Y75, (2) a para-binding-pocket cluster, centered around an important proline residue (P17) facilitates movement, (3) a lower cytoplasmic-oriented cluster, containing highly conserved residues D60, D288, and P289, and (4) a distal aromatic cluster of four cyclic-side-chain-containing amino acids (F146, F150, F184, and Y188). The last four residues are involved in intra- and inter-α-helical hydrophobic (ring–ring) interactions and serve as stabilizing forces for TMIV and TMV.[71] This stabilization is important for

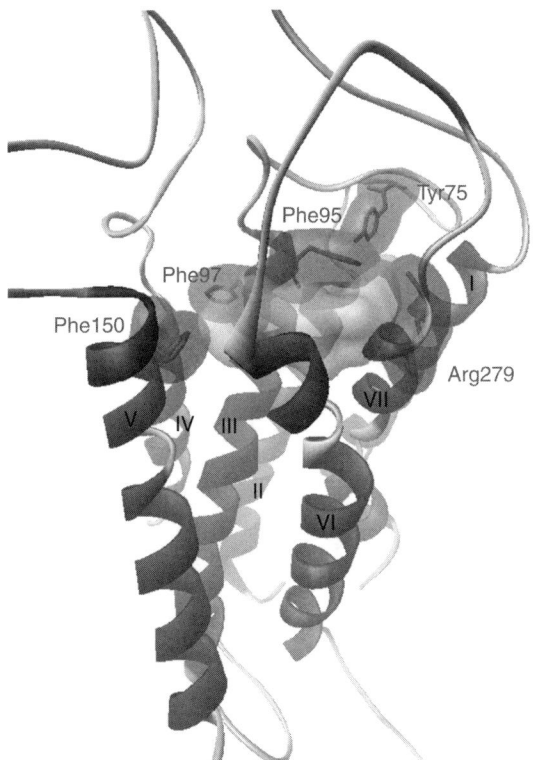

FIG. 4. Diagram showing critical residues involved in prostacyclin binding. Within the binding pocket is a model of prostacyclin. Critical residue interactions as described in the text are shown within their respective transmembrane domains in roman numerals.

achieving an active receptor conformation. Furthermore, Stitham et al.[72] targeted nine TM proline residues within the hIP receptor using site-directed (alanine-scanning) mutagenesis to assess residues necessary for receptor binding and activation. The P179A (TMV) mutation affected binding but not activation. P17A (TMI), P69A (TMII), P89A (TMIII), and P289A (TMVII) mutants affected activation but not agonist binding, while the P154A (TMIV) and P254A (TMVI) affected both binding and activation. The P141A (TMIV) and P285A (TMVII) affected neither binding nor activation.[72] The P89A (TMIII) mutation demonstrated wild-type binding, but suffered a 10-fold reduction in efficacy owing to a marked reduction in activation. This would suggest P89 is important in α-helical movement. Indeed, movement of TMIII has been found to be important in activation of rhodopsin, another member of this family of receptors.[73] These findings support a model in which proline residues act as molecular swivels or hinges allowing the receptor to attain conformations necessary for ligand binding and subsequent activation.[72]

D. Cysteines and Palmitoylation–Isoprenylation

The cytoplasmic region of the hIP receptor contains some typical Class A features, including the highly conserved ERY/C (Glu-Arg-Tyr/Cys) motif at the junction between TMIII and the second cytoplasmic loop (Fig. 2), implicated in the regulation of receptor conformational states and G-protein activation.[74] A highly conserved palmitoylation–isoprenylation (CAAX) sequence, CCLC (Cys-Cys-Leu-Cys), provides a site for lipid anchoring, which tethers part of the C-terminal tail to the cytoplasmic membrane, forming a fourth cytoplasmic loop. Work by Miggin *et al.* established this region to be formed by palmitoylation at the first and last Cys residue (i.e., C308 and C311).[41] Differential mutagenesis of this motif showed that either C308 or C311, alone, is sufficient for hIP coupling to Gαs, but C308 is specifically needed for Gαq coupling. More recent work by O'Meara and Kinsella, investigating the effect of SCH66336 (farnesyl protein transferase inhibitor) on effector coupling and intracellular signaling by hIP, revealed evidence supporting that this C-terminal isoprenylation motif (CSLC) is the target of farnesylation (posttranslational addition of 15-carbon isoprene polymer), catalyzed by the farnesyl protein transferase enzyme (FTase). These studies demonstrate that the level of cicaprost-mediated cAMP generation and cicaprost-induced receptor internalization were both significantly reduced in the presence of SCH66336 compared with vehicle-treated cells, indicating that: (1) the hIP receptor is farnesylated at the C-terminal CSLC motif, (2) this modification is required for proper receptor coupling and activation, and (3) the hIP is a direct target of SCH66336 *in vitro*.[75]

E. Serines and Receptor Regulation

A number of C-terminally located serine residues (e.g., S328 and S374) are thought to be phosphorylated by GPCR kinases (GRKs) or second messenger-activated kinases (PKC and PKA), and play potential roles in either agonist-induced phosphorylation and/or kinase-mediated receptor desensitization.[64] In murine models, it has been demonstrated that cAMP-dependent activation of PKA, presumably via Gαs-mediated signaling, leads to the phosphorylation of S357 within the C-terminal of the mouse IP (mIP) receptor. Such phosphorylation couples mIP with Gαi, leading to the inhibition of AC as well as Gαq-mediated activation of phospholipase C (PLC).[76] Such evidence supports a model where stimulation of mIP and Gαs can lead to subsequent phosphorylation events that favor coupling to noncanonical signaling cascades such as Gαi and Gαq.

As mentioned earlier, alternative IP coupling to multiple G-protein–effector signaling pathways has been shown to occur, but in a cell-type- and perhaps species-dependent manner. Subsequent work by Lawler and Kinsella found that such alternative coupling–signaling events were not reproducible across species, particularly for hIP. They showed that HEK-expressed mIP and hIP coupled to Gαq, leading to PLC activation. However, unlike mIP, hIP did not couple to Gαi.[64] Moreover, this work also demonstrated that hIP independently couples to Gαq, while mIP depends on PKA phosphorylation for Gαs to Gαq switching. Computation analysis identified residue S357, within the C-terminal consensus site RPASGRR (Arg-Pro-Ala-Ser-Gly-Arg-Arg), as the target for PKA phosphorylation in the mIP receptor. Interestingly, this sequence is not found in the hIP receptor, which contains a site for PKC phosphorylation at residue S328, QLASGRR (Glu-Leu-Ala-Ser^{328}-Gly-Arg-Arg).[64] Thus, it is clear that important differences in cell- and species-specific expression systems may be subtle, as with minor changes in protein sequence, but can have significant consequences in receptor–effector coupling. It is also clear that hIP signal transduction is not confined to a single linear signaling pathway, involving instead a variety of signaling systems that interact in complex patterns and permutations.

F. Model of Receptor-Bound Prostacyclin (PGI$_2$) Ligand

Stitham et al. created a three-dimensional PGI$_2$ molecule, based on the known chemical structure, as well as a previously predicted conformation of a receptor associated PGI$_2$.[42] Initial modeling constraints required the adjustment of individual atoms such that they would conform to the chemically acceptable limits of the ligand structure. This included the adjustment of all covalent bonds to agree with standard lengths and angles: sp^3-hybridized C–C single bond = 1.54 Å and 109.5°; sp^2-hybridized C=C double bond = 1.33 Å and 120.0°; sp^3-hybridized C–O single bond = 1.40 Å and 108.0°; and sp^2-hybridized C=O double bond = 1.20 Å and 120.0°. Because of substantial torsional strain, all interior angles for both five-membered rings were approximated at 108.0° with the exception of the oxane ring C–O–C bond angle, set closer to 112.0°. The root mean square deviation for bonds and angles were calculated as 0.006 Å and 0.4°, respectively, compared with the standard numbers referenced earlier. The two rings were constructed to assume a common envelope configuration with the characteristic four coplanar atoms combined with a fifth member out of plane. The structural conformation of the receptor-bound prostacyclin molecule is similar to that of previous investigations in modeling conformations of receptor-associated PGI$_2$, where the α-chain is maintained in a bent conformation (back upon the two centralized rings), whereas the long hydrophobic ω-chain is in an extended configuration. This configuration was independently confirmed using MacSpartan Pro software (Wavefunction, Inc., Irvine, CA) where the energy-minimized conformations (-7 to -10 kcal/mol) all exhibited such a ring and α-chain conformation with the major variances being in position of the ω-chain.

G. NMR-Based Structures of the hIP Receptor

While site-directed mutagenesis and naturally occurring mutations provide valuable information regarding modification of structural interactions, biophysical studies provide a direct means of assessing the validity of structural models. Though relatively little biophysical information is available concerning the overall structure of the prostacyclin receptor, considerable structural data may be collected through the approach of dividing the larger polypeptide into smaller more manageable pieces to enable direct biophysical measurements that would otherwise be unapproachable. Zhang and colleagues applied such a constrained peptide approach to the first intracellular loop of the human prostacyclin receptor, using solution NMR analysis, to investigate the potential interaction of this segment with G-protein (i.e., binding of Gαs).[77] The resulting structure for the constrained peptide loop, in the absence of Gαs, measured a separation of 10.0 Å between the α-carbon of residue L38 and L52, providing indirect biophysical support to our own calculated distance of 10.1 Å between these atoms in the rhodopsin-based homology model (Stitham, Gleim et al., unpublished data). Comparing results from both approaches (NMR and homology model), in the absence of Gαs, the NMR solution structure for the first intracellular loop of the hIP maintains the same relative conformation as the homology model for residues within this region. Most notably, both models demonstrate the side chain of R45 extending away from the center of the loop. In the homology model of Stitham and Gleim, this residue participates in hydrogen bonding with E234, and the backbone carbonyl groups of L126 and V235 (Stitham, Gleim et al., unpublished data).

Similar NMR and homology studies have been performed on the extracellular loops of the hIP receptor, owing to their proposed involvement (particularly the second loop) in ligand recognition.[78] Considerable deviation is observed between the two models in this region; however, this difference is likely due to the addition of a secondary disulfide bond between C5 (on the N terminus) and C165 (within this loop). Incorporation of this disulfide may further stabilize the receptor, allowing the N-terminal chain to lay across the extracellular receptor surface, bridging the second and third extracellular loops through the combination of disulfide formation and hydrogen bonding of R6 to D267.

Based on previous findings suggesting the importance of R212 in receptor activation,[40] the third cytoplasmic loop is of considerable interest. Unfortunately, no biophysical or chimeric data is available to compare with our investigations into this region. Nevertheless, secondary structure prediction of the overall polypeptide suggests a considerable degree of helicity in the region containing this residue. Minimized structures containing the predicted helical formation in this region suggest that R212 may be capable of forming hydrogen bonds with Q216 and the side-chain hydroxyl (–OH) of S205. Nearby R219, also contained in the predicted cytoplasmic helix, would similarly be capable of forming a potentially important hydrogen bond with S122 on the second intracellular loop (Stitham, Gleim et al., unpublished data).

III. Pathophysiology

A. Atherothrombosis

1. VSMC Phenotypic Modulation and Implications for Disease

VSMCs play a critical role in the development of atherosclerosis. Phenotypic modulation of VSMC is critical in pathophysiologic processes and conditions, such as vasculogenesis and angiogenesis, atherosclerosis, restenosis, and hypertension.[79] Unlike cardiac myocytes, VSMC do not terminally differentiate, but retain the plasticity to dedifferentiate to a synthetic phenotype in response to growth factor stimulation during vascular injury or stress.[80] In a normal, mature blood vessel, VSMC exhibits a differentiated "contractile" phenotype, characterized by expression of smooth muscle (SM)-specific contractile markers such as SM myosin heavy chain (SM-MHC), SM α-actin, h-caldesmon, and calponin, each important for regulating the main function of these cells: contraction.[81] Upon changes in local environmental signals, VSMC dedifferentiate and reenter the cell cycle, characterized by increased proliferation and migration rates, increased synthesis of extracellular matrix (ECM) components, and decreased expression of SM-specific contractile markers.[79,82] In the past, proliferation and differentiation were assumed to be obligately coupled, with loss of SM differentiation markers a prerequisite for VSMC proliferation. It is now understood that these processes are not mutually exclusive. Growth factors affect VSMC proliferation independent of differentiation marker expression and vice versa.[80,83–86] Furthermore, during late embryogenesis, VSMC exhibits high proliferation rates with high expression levels of SM-specific markers.[87,88] Additionally, SM cell (SMC) in advanced atherosclerotic lesions show reduced rates of proliferation and low expression of SM-specific differentiation markers.[89,90]

The vascular endothelium- and endothelial-derived secreted factors have been implicated in modulating VSMC phenotype.[91,92] The idea that the endothelium is an endocrine–paracrine organ and not just an inert lining of the blood vessel has only been appreciated within the past two decades, particularly in response to the discovery of vasoactive substances such as prostacyclin,[93,94] NO,[95–97] endothelin-1 (ET-1),[98–100] and platelet-activating factor (PAF).[101] As a single-cell layer covering the internal surface of the vessel, the endothelium is directly exposed to blood flow and can directly sense changes in hemodynamic forces and respond accordingly by releasing these vasoactive substances.[102] Thus, it is not surprising that endothelial dysfunction is associated with altered vascular growth and dysregulated vascular remodeling, which occur in atherosclerosis and hypertension. This is likely due to the loss of endothelial-derived factors such as prostacyclin (PGI_2) and NO.[93–96] Loss of such factors, along with increased exposure to inflammatory stimuli and growth factors, contributes to VSMC proliferation, migration, and dedifferentiation from contractile to synthetic phenotype, leading to intimal thickening of the affected artery.[102]

2. Role for Prostacyclin in Promoting VSMC Differentiation

Although it is known that prostacyclin receptor signaling inhibits VSMC proliferation and migration, the Martin laboratory has demonstrated that prostacyclin may also protect the vasculature by maintaining VSMC in a differentiated state through induction of SM-specific genes characteristic of the contractile phenotype. This would maintain the contractile function, while preventing detrimental effects of dedifferentiation, such as lumen loss due to VMSC migration, proliferation, and matrix deposition. The prostacyclin mimetic iloprost-induced dedifferentiated adult human VSMC to assume a contractile differentiated phenotype, increasing expression of SM-specific differentiation markers as well as changing morphology.[49] Furthermore, induction of SM-specific markers required activation of the prostacyclin receptor–effector pathway Gs/AC/cAMP/PKA, as a cAMP analog was sufficient to induce differentiation markers. PKA activation is necessary for IP-induced VSMC differentiation.[49] It has been well established that cAMP triggers cellular differentiation in the pheochromocytoma cell line PC12.[103,104] This cAMP-dependent neural cell differentiation is mediated by the induction of a set of specific genes via activation of the transcription factor CREB by PKA.[104] Interestingly, IP regulates SM-specific differentiation markers at the level of transcription.[49]

Prostacyclin receptor induction of VSMC differentiation highlights another important role in vasculature protection and provides a potential mechanism for how differentiation is modulated under physiological conditions. Importantly, loss of prostacyclin synthesis due to endothelial dysfunction may allow the VSMC dedifferentiation, proliferation, and migration that contribute to pathophysiological conditions such as restenosis, atherosclerosis, and blood vessel remodeling.

3. MECHANISMS OF PROSTACYCLIN-INDUCED ANTIPROLIFERATIVE EFFECTS IN VSMC

Several studies provide compelling evidence that prostacyclin inhibits VSMC proliferation both *in vitro* and *in vivo*. Genetic deletion of the prostacyclin receptor in mice resulted in greater injury-induced VSMC proliferation compared to wild-type counterparts, contributing to neointimal formation.[105] Conversely, overexpression of prostacyclin synthase inhibits VSMC proliferation *in vitro* and neointimal formation *in vivo*.[106,107] Additional studies have demonstrated PGI_2 analogs to inhibit VSMC proliferation, both *in vitro* and *in vivo*, by inhibiting G1- to S-phase cell cycle progression.[108,109] G1- to S-phase progression is largely modulated by the activities of G1-phase cyclin-dependent kinases (cdks), particularly cyclin D-cdk4/6 and cyclin E-cdk2.[110] Once activated, these complexes phosphorylate the pocket protein family members, Rb, p107, and p130, responsible for modulating E2F-dependent gene transcription, such as cyclin A.[111] To understand the cell cycle influences of prostacyclin signaling, Kothapalli *et al.* examined the effect of the specific IP agonist, cicaprost, on the rat A10 SM cell line and on primary murine VSMC.[108] The authors found that Rb and p107 were hypophosphorylated in response to cicaprost and this was associated with decreased cyclin A promoter activity and inhibition of cell cycle progression by cicaprost was due to a selective inhibition of the activity of cyclin E-cdk2 but not cyclin D-cdk4/6.[108]

Interestingly, in mouse embryonic fibroblasts (MEFs)[109] and canine VSMC,[112] prostacyclin analogs cicaprost,[109] and beraprost sodium (BPS),[112] respectively, block downregulation of a cyclin-dependent kinase inhibitor, p27kip, in the presence of mitogenic stimuli. It had been previously shown that p27kip inhibits cell cycle progression by keeping cyclin E-cdk2 in an inactive state.[113] Ii *et al.* showed that BPS prevented serum-induced downregulation of p27kip in cultured canine VSMC,[112] and inhibited neointimal formation following carotid balloon injury in part by maintaining p27kip expression. Effects on p27kip appear to be via Gs activation, as the effect is mimicked by a cAMP analog and opposed by AC inhibitor 2′,5′-dideoxyadenosine (DDA).[112]

Cicaprost stabilization of p27kip is necessary for prostacyclin inhibition of G1- to S-phase progression.[109] In MEFs null for p27kip, cicaprost no longer blocked S-phase entry, cyclin A induction, or cyclin E-cdk2 activation. Furthermore, cicaprost inhibited serum-induced expression of Skp2, the substrate-targeting component of the E3 ubiquitin ligase complex responsible for p27kip degradation.[114–116] Expression of exogenous Skp2 rescued the p27kip downregulation and S-phase progression in MEFs treated with cicaprost, suggesting that prostacyclin regulates p27kip through modulated expression of Skp2. Interestingly, the cAMP elevating agents, 8-Br-cAMP and forskolin, stabilized p27kip and inhibited G1- to S-phase progression without affecting Skp2 levels,[109] suggesting that cicaprost-induced inhibition of Skp2 gene expression is not cAMP-mediated.

Ii et al. and Stewart et al. independently demonstrated that PGI_2 signaling regulates p27kip and that this plays a critical role in PGI_2-induced inhibition of proliferation.[109,112] However, Stewart and colleagues demonstrated that cicaprost regulation of Skp2, and thus regulation of p27kip, was independent of cAMP generation, whereas Ii et al. showed that the BPS regulation of p27kip was cAMP-dependent.[109,112] Together, these studies suggest that PGI_2 may inhibit proliferation through both cAMP-dependent and cAMP-independent mechanisms.

VSMC proliferation is a major factor in the development of vascular disease, and regulation of p27kip expression is an important mechanism by which prostacyclin signaling can inhibit cellular proliferation. Notably, it has been shown that p27kip is critical in vascular injury: p27kip expression in SMC decreases following balloon injury[117] and overexpression of p27kip in the vessel wall can inhibit injury-induced neointimal formation.[118,119] Furthermore, as suggested by Stewart et al., inhibition of Skp2 expression seems to be a critical event responsible for the antimitogenic effect of prostacyclin. Thus, Skp2 may be a potential therapeutic target for CVD.[109]

4. Mechanisms of Prostacyclin Inhibition of VSMC Migration

IP agonist inhibits PDGF-BB-induced migration of cultured human aortic VSMC.[120] In recent studies, Bulin et al. proposed that disruption of focal adhesion formation and inhibition of focal adhesion kinase (FAK) may mediate IP inhibition of migration in human aortic VSMC.[121] Interestingly, these effects required Gs activation of the cAMP/PKA pathway, as the cAMP analog db-cAMP and the AC activator forskolin mimicked the effect of IP, whereas the PKA inhibitor H89 reduced these effects.[121] The authors speculated that hypophosphorylation of FAK may be due to IP-induced activation of protein tyrosine phosphatases (PTPs) based on evidence that PTP inhibitors, vanadate and calpeptin, prevent IP inhibition of FAK phosphorylation.[121]

5. Other Roles in Atherosclerosis

Atherosclerosis is a complex disease characterized by inflammation, lipid accumulation, cell death, and fibrosis.[122] It is the leading cause of mortality in many countries, including the USA. The first phase of this disease is characterized by accumulation and penetration of monocytes into the intimal layer of bloods vessels, followed by proliferation and migration of VSMCs. Over time, lesions mature and can lead to vessel occlusions that limit blood flow, or in some cases rupture, resulting in thrombotic occlusion of the vessel and subsequent myocardial infarction (MI) and stroke. Prostacyclin plays many roles in VSMCs and platelets relevant to preventing the progression of these events. Several reports have described increased COX-2 expression in human atherosclerotic lesions[123–125] suggesting a role for COX-2 in atherosclerosis. Concurrently, increased production of prostaglandins was described in association with atherosclerosis.[13,126,127] The roles of the prostaglandins in atherosclerosis appear to be antagonistic: TXA_2 and prostaglandin E_2 (PGE_2) have an atherogenic effect, whereas PGI_2 and PGD are antiatherogenic. TXA_2 is a vasoconstrictor and smooth muscle cell mitogen produced by platelets and monocytes.[128,129] Macrophage PGE_2 synthesis can promote atherosclerotic plaque instability through PGE_2-dependent metalloproteinases (MMP-2 and MMP-9),[127,130] suggesting that the balance between PGE_2 and PGD synthases influences plaque instability. There appears to be a similar balance between thromboxane and prostacyclin.[12] IP, a stable analog of prostacyclin, downregulates lymphocyte adhesion to endothelial cells, potentially inhibiting early events of atherosclerosis.[131] The ability of HDL to induce COX-2 expression and PGI_2 production by both endothelial and SMCs has been widely demonstrated[32,132,133] supporting the antiatherogenic response by PGI_2.[134] Notably, PGI_2, but not PGE_2, induces cholesteryl ester hydrolase activity, whereas PGE_2, but not PGI_2, inhibits Acyl-CoA cholesterol acyl-transferase activity in human vascular SMC. Cholesteryl ester hydrolase catalyzes the first step in the removal of cholesterol from foam cells whereas Acyl-CoA cholesterol acyl-transferase, which modifies cholesterol in such a way that it can be stored in cells, plays a key role in the formation of cholesterol-loaded macrophages/foam cells.[135–137]

6. Thrombosis

The prostacyclin receptor is abundant on platelets and VSMCs. As early as 1976, it was postulated that a balance between antiaggregatory (prostacyclin) and proaggregatory (thromboxane) substances contributed to maintaining the integrity of vascular endothelium and thrombus formation.[4] This was further reinforced by the first prostacyclin receptor knockout mouse which exhibited increased susceptibility to thrombosis in addition to a reduced inflammatory and pain response.[30] Clinical trials such as the VIGOR study group similarly showed rofecoxib treatment (selective COX-2 inhibitor), in rheumatoid arthritis patients, reduced GI side effects compared to naproxen but increased incidence of thrombotic arterial events (MI).[37] Clearly, prostacyclin limits the response of thromboxane, as elegantly demonstrated with receptor knockout mice studies. The cardiovascular events associated with selective COX-2 inhibitors may be associated with inhibition of prostacyclin and not thromboxane.[39]

B. Role of hIP in Inflammation and Pain

In addition to its function maintaining vascular homeostasis,[138] prostacyclin plays a prominent role in pain perception and inflammation. Nonsteroidal anti-inflammatory drugs (NSAIDs) known for analgesic, anti-inflammatory, and antipyrogenic effects, work by inhibiting prostanoid synthesis.[139] Prostacyclin signaling in particular contributes to inflammatory symptoms. Studies on receptor localization and downstream signaling elements of prostacyclin have revealed further details regarding pain perception.

Prostacyclin receptor induction of inflammatory pain was highlighted in a 1997 study by Murata and colleagues on knockout mice.[30] Prostacyclin had been previously linked to inflammation due to its stimulation of vasodilation.[138,140] Murata and his colleagues deleted both copies of the IP gene from mice.[30] Under conditions stimulating inflammation and pain response, prostacyclin receptor knockout mice showed reduced edema, exudate volume, and acetic acid-induced writhing compared to those observed in wild-type mice. Knockout mice responded to inflammatory and pain stimuli at levels similar to those in wild-type mice that had reduced levels of prostacyclin synthesis due to pretreatment with the aspirin-like NSAID indomethacin.[30] This study implicated prostacyclin signaling in specific symptoms associated with inflammation.

Prostacyclin hyperalgesia has been examined in the context of receptor localization in neurons and downstream events in the signaling pathway. The prostacyclin receptor can be found on sensory neurons, with the greatest density on primary afferent nerve terminals.[141] Mice show writhing behavior in response to prostacyclin analog injection, supporting prostacyclin activation of primary afferents.[142–145] Following administration of prostacyclin and the prostacyclin analog cicaprost, rats demonstrate increased activity of articular mechanonociceptors,[145,146] suggesting recruitment of pain-sensing neurons.

Facilitating pain perception has been attributed in part to specific downstream signaling factors, including PKA and PKC. Upon binding the prostacyclin ligand, the prostacyclin receptor signals to adenylate cyclase catalyzing cAMP formation, activating PKA. Mice with a null mutation for the Type 1β regulatory subunit of PKA show diminished nociceptive inflammatory response, supporting PKA's role in initiating sensory neuron sensitization.[147,148] Although prostacyclin does not directly activate PKC, PKA- and PKC-dependent pathways have been hypothesized to interact with one another, and may synergistically contribute to neuron sensitization.[149] Similar to those mice with dysfunctional PKA, transgenic mice lacking the γ isoform of PKC also display reduced inflammatory nociceptive responses.[150,151]

Downstream events in the prostacyclin signaling pathway may also contribute to pain perception by altering neuropeptide release and ion channel activity. Increased cAMP concentration, via prostacyclin, potentiates neuropeptide release from dorsal root ganglion cultures.[152] Prostacyclin analog carbacyclin suppresses outward K^+ current in rat dorsal root ganglion.[153,154] In guinea pig nodose ganglion, prostacyclin inhibits a slow after-hyperpolarization K^+ current,[155] and in cultured dorsal root ganglion cells prostacyclin enhances a capsaicin-activated inward current.[156]

The ability to modulate ion channel activity, and hence neuronal excitation, suggests a potent role for prostacyclin in the nervous system,[141] and supports the prominence of prostacyclin inflammatory nociception. Animals lacking either the prostacyclin receptor or any of the various regulatory subunits in downstream components of the signaling pathway show diminished inflammation and pain response compared to wild-type controls. While the role of prostacyclin signaling in inflammation and pain suggests that this receptor might be a potent antinociceptive target,[141] the importance of prostacyclin signaling in cardioprotection[30,39,157] poses challenges to this method of pain management.

C. Role of hIP in Parturition

Prostaglandins are known to regulate myometrial stimulation.[158,159] Prostaglandin $F_2\alpha$ ($PGF_{2\alpha}$) and, to a lesser extent, PGE_2, directly stimulate uterine contractility *in vitro*.[160–163] In human pregnancy, prostaglandins are produced by the amnion and chorion, decidua, and myometrium,[158] with prostacyclin (PGI_2) being the most highly synthesized prostaglandin by the pregnant human myometrium.[164–167] This appears paradoxical as synthesis is increased prior to and during labor,[165,168,169] despite prostacyclin being a potent SM relaxant.[170] The relaxing effects of PGI_2 have been attributed to its activation of the cAMP/PKA pathway.

Many potential physiologic roles for the relaxant effects of prostacyclin during labor have been proposed, including the prevention of myometrial cell overstimulation during labor,[171] and promoting cervical softening.[159] It has also been suggested that because prostacyclin has potent vasodilatory[7,172] and antiplatelet activity,[173] it may be important for preventing intrapartum thromboembolic complications in the maternal and fetoplacental circulation.[169] Others suggest that prostacyclin may actually promote myometrial contractility instead of relaxation. Previous studies in tissue strips from nonpregnant[174] and pregnant[175] rat myometrium demonstrated that acute prostacyclin exposure leads to contractile activity. However, results from pregnant human myometrial tissue strips have been conflicting, showing either an inhibitory[166,176] or a biphasic effect[176,177] of acute prostacyclin or prostacyclin analog treatment on contractions. Fetalvero and colleagues, recently, demonstrated that prostacyclin signaling through PKA upregulates connexin 43 and contractile protein expression, enhancing the contractile response to oxytocin in pregnant human myometrial tissue.[53] This has implications for endogenous prostacyclin as an important factor in regulating myometrial activation, a critical step in the initiation and progression of parturition. This work suggests intriguing avenues for further investigation, as prostacyclin antagonists may be therapeutic in treatment of preterm labor, with agonists potentially useful in promoting efficient labor.

IV. Therapeutics

We will now focus on the effects of therapeutics (coxibs and NSAIDs) on prostacyclin and its receptor, in addition to the prostacyclin receptor as a therapeutic target (receptor agonists and antagonists). Given the clinical implications, it is urgent to address these very important issues. NSAIDs are commonly used amongst high cardiovascular risk patients who have concurrent arthritis. This could potentially accelerate CVD progression. Furthermore, although currently used effectively in the treatment of pulmonary hypertension, prostacyclin agonists could be a new therapeutic target in the treatment of atherothrombosis.

A. Selective COX-2 Inhibition (Coxibs)

Perhaps, the most dramatic example linking PGI_2 signaling and atherogenesis is the recent withdrawal of selective COX-2 inhibitors. Fitzgerald and others had raised the concern early on that inhibition of COX-2 might predispose patients to adverse coronary events.[33] Both celecoxib (Celebrex^{TM}) and rofecoxib (Vioxx^{TM}) suppressed urinary 2,3-dinor-6-keto $PGF_{1\alpha}$, similar to traditional NSAIDs (tNSAIDs), indicating inhibition of PGI_2 production,[33,34] disrupting the balance between TXA_2 and PGI_2 signaling in favor of the atherothrombotic effects of TXA_2.

Given this evidence, it would be reasonable to hypothesize those patients with underlying CVD and those with increased risk factors for CVD such as hypertension, diabetes, family history, age, and smoking history would be at greater risk for adverse cardiovascular events with COX-2 inhibition and the resulting reduction of prostacyclin production. The VIGOR[37] and APPROVe[46] trials have since confirmed this hypothesis, precipitating the withdrawal of select drugs in the class.[58] The VIGOR study examined the protective effect of rofecoxib (Vioxx^{TM}) on catastrophic GI side effects relative to tNSAIDs and found a twofold reduction in the incidence of serious GI side effects with rofecoxib versus the tNSAID naproxen, but at the cost of a fivefold increase in the incidence of MI.[37] The coronary artery bypass graft (CABG) study examined valdecoxib (Bextra^{TM}) and its prodrug parecoxib (Dynastat^{TM}) compared to placebo for the treatment of postoperative pain following CABG.[47] This study found a nearly fourfold increase in risk of MI in the valdecoxib-treated group compared to those treated with placebo. The AP-PROVe trial[46] found a divergence in the treatment groups to be significant at 18 months compared to the CABG trial, which found a significant difference in adverse CVD treatment outcomes at 4 days for the parecoxib/valdecoxib group compared to those receiving placebo. These findings suggest that in the context of vascular injury, as in patients undergoing CABG, atherothrombosis is accelerated. Therefore, it would appear that the balance of IP and TP signal transduction and the resulting effects on cardiovascular health are mitigated by factors known to cause vascular damage and may extend to factors known to aggravate vascular disease such as smoking, diabetes, and hypertension.

B. COX-1/2 Inhibitors (NSAIDs)

NSAIDs are a chemically heterogeneous group of drugs that are commonly used in the general population for treating pain and inflammatory conditions. They act through inhibition of COX-1 and COX-2 and are thus distinguished on the basis of their COX-isoenzyme selectivity (Table I). Concerns that such drugs may potentially lead to acceleration of CVD were brought to the forefront with the discovery that selective COX-2 inhibitors such as rofecoxib and celecoxib could lead to increased cardiovascular events such as heart attacks.[58,59] There is increasing concern that other NSAIDs commonly used in the general population (e.g., ibuprofen, indomethacin) may have similar effects.

TABLE I
TABLE OF SELECTED NSAIDs AND THEIR SELECTIVITY FOR THE COX ISOENZYMES

	COX-1 IC50 (μM)	COX-2 IC50 (μM)	References
Aspirin	410	61	191
Indomethacin	1.67	24.6	192
Phenylbutazone	16.0	>100	193
Ibuprofen	2.6	1.53	192
Diclofenac	0.9	1.5	192
Nimesulide	70.0	1.27	192
NS-398	75.0	1.77	192

References are provided from where the information was obtained.

C. Pulmonary Hypertension

Prostacyclin analogs are effectively used in the treatment of pulmonary hypertension,[178] a fatal illness characterized by increased pulmonary mean arterial pressures secondary to pulmonary vasoconstriction and marked by VSMC proliferation, remodeling of the pulmonary vessel wall, and thrombosis.[179–181] Vascular-wall remodeling is an important feature of this disease and has been attributed, at least in part, to an imbalance in the ratio of prostacyclin to thromboxane.[18] Intravenous infusion of the prostacyclin analog epoprostenol (Flolan[TM]) has been a recommended therapy since the mid-1990s[27,182] and has been shown to increase exercise tolerance and improve hemodynamics in patients. However, due to the short half-life and low pH instability of epoprostenol, it must be delivered via pump and is associated with a number of side effects such as catheter-related sepsis.[183] As an alternative, new prostacyclin analogs have been developed, including subcutaneous treprostinil (Remodulin[TM]),[184] oral beraprost (Dorner[TM]),[185] and inhaled iloprost (Ventavis[TM]).[35,57] It remains to be seen whether prostacyclin analogs are useful for diseases in other vascular beds such as the coronary arteries. Its known actions both on VSMCs and platelets suggest that it would be beneficial in patients with atherothrombosis.

V. Genetic Variants

The study of genetic variants and their influence on disease, disease progress and therapy is becoming increasingly appreciated in this postgenomic era. Currently, there are over 140,000 articles with the word "polymorphism" either in abstract or title (PubMed search). It is anticipated that over 15 million polymorphisms will be detected in the human genome. As described earlier, the evidence for the important influence of prostacyclin on the cardiovascular system has led to intense studies on genetic variants of the hIP receptor. Stitham and his colleagues were the first to describe and characterize single-nucleotide polymorphisms (SNPs) in the human prostacyclin receptor.[40] They determined that an R212H receptor variant, in the critical third cytoplasmic loop, impairs receptor function. Of particular interest for clinical consequences, defects in the structure and function of the polymorphic R212H receptor were more pronounced under stress conditions such as acidosis, for example, pH 6.8.[40] During pathological stress such as respiratory or cardiac failure (where severe acidosis may provoke defective prostacyclin binding), the added receptor defect may allow for unchecked vasoconstriction and/or thrombosis, warranting urgent correction of pH to restore proper receptor function. Since these initial observations, two additional genetic variants in the coding region of *PTGIR* were described in a Japanese cohort,[186] with 27 other genetic variants identified in a larger pharmacogenetic study through sequencing 1455 human samples[187] (Table II).

The importance of the prostacyclin receptor in protection against CVD has been, recently, highlighted in two independent populations. An R212C variant was found, at the same amino acid position as R212H, to accelerate CVD.[51,52] However, in the absence of significant risk factors there appears to be no significant R212C phenotype.[51] This finding parallels and reinforces the influence of COX-2 inhibition, where reduction in prostacyclin production appears to increase cardiovascular events in high cardiovascular risk patients, but lacks association in other study populations. This important principle illustrates a broader phenomenon in which many high profile manuscripts reporting polymorphism–disease associations are not uniformly confirmed by subsequent studies. For such association studies, the role of the variant in separate populations with differential risk factor profiles may be a critical factor. We are also rapidly becoming aware that genetic variants often remain silent under normal physiologic conditions, or throughout childhood and early adulthood, with functional abnormalities becoming apparent only in the diseased state, or under pathophysiological insult.[188–190] For the prostacyclin signaling pathway, this was demonstrated most dramatically with the prostacyclin receptor knockout mice, where underlying stress or injury was required to reveal that the lack of this receptor promotes increased thrombosis or intimal hyperplasia.[30,39,157]

The mechanism for accelerated atherothrombosis appears to be largely due to a combination of reduced thrombosis inhibition from reduced cAMP signaling in platelets, and an inability to reduce human coronary VSMC proliferation and dedifferentiation,[49] also from reduced cAMP production. For the R212C variant this appears to arise from disruption of a critical interaction between R212 and S205, required for stabilizing conformation in the critical G-protein interacting third intracellular loop. A decrease in the effective dose of prostacyclin, due either to receptor signaling defects or to inhibition of prostacyclin production (COX-2 inhibition), would similarly enable thrombosis and promote dedifferentiation and proliferation in human coronary VSMCs. Aspirin (predominant COX-1 inhibitor) itself appears insufficient to oppose the atherothrombotic state in the R212C patients, as 90% of the coronary heart disease R212C patients were on low-dose aspirin therapy. It remains to be seen whether other hIP genetic variants will also influence CVD. Furthermore, the influence of haplotypes, including the influence of thromboxane receptor genetic variants on genetic variants of prostacyclin, is now required. The solutions are likely to be very complex, necessitating collaborative efforts with larger populations. It also remains unknown whether these dysfunctional genetic variants influence other processes such as pain, inflammation, and parturition.

TABLE II
Genetic Variants Found from Sequencing 1455 Genomic Samples[187]

	Nucleotide change
V15A	T → C
V25M	G → A
G27G	T → A
L33L	G → T
G36G	C → T
P43P	G → A
F49F	C → T
V53V	G → C
P69P	G → A
R77C	C → T
F102F	C → T
L104R	T → G
M113T	T → C
G181A	G → C
L186L	G → A
R212C	C → T
R212H	G → A
R212R	C → T
R215C	C → T
P226T	C → A
G231R	G → A
R279C	C → T
P289P	C → T
I293N	T → A
S319W	C → G
S319L	C → T
S328S	A → C
E354D	G → T
S369R	C → A
T373T	G → A
S374S	G → A

Italics show the nonsynonomous mutations.

It is now clear that we are only exploring the tip of the iceberg and extensive sequencing, and characterizations are necessary to determine the roles these genetic variants play not only in the cardiovascular system but also with pain, inflammation, and parturition.

References

1. Gryglewski R, Vane JR. The inactivation of noradrenaline and isoprenaline in dogs. *Br J Pharmacol* 1970;**39**:573–84.
2. Gryglewski R, Vane JR. Rabbit-aorta contracting substance (RCS) may be a prostaglandin precursor. *Br J Pharmacol* 1971;**43**:420P–1P.
3. Bunting S, Gryglewski R, Moncada S, Vane JR. Arterial walls generate from prostaglandin endoperoxides a substance (prostaglandin X) which relaxes strips of mesenteric and coeliac ateries and inhibits platelet aggregation. *Prostaglandins* 1976;**12**:897–913.
4. Moncada S, Gryglewski R, Bunting S, Vane JR. An enzyme isolated from arteries transforms prostaglandin endoperoxides to an unstable substance that inhibits platelet aggregation. *Nature* 1976;**263**:663–5.
5. Gryglewski RJ. Prostacyclin among prostanoids. *Pharmacol Rep* 2008;**60**:3–11.
6. Whittaker N, Bunting S, Salmon J, Moncada S, Vane JR, Johnson RA, et al. The chemical structure of prostaglandin X (prostacyclin). *Prostaglandins* 1976;**12**:915–28.
7. Dusting GJ, Moncada S, Vane JR. Prostacyclin (PGX) is the endogenous metabolite responsible for relaxation of coronary arteries induced by arachindonic acid. *Prostaglandins* 1977;**13**:3–15.
8. Moncada S, Higgs EA, Vane JR. Human arterial and venous tissues generate prostacyclin (prostaglandin X), a potent inhibitor of platelet aggregation. *Lancet* 1977;**1**:18–20.
9. Moncada S, Herman AG, Higgs EA, Vane JR. Differential formation of prostacyclin (PGX or PGI2) by layers of the arterial wall. An explanation for the anti-thrombotic properties of vascular endothelium. *Thromb Res* 1977;**11**:323–44.
10. Gorman RR, Bunting S, Miller OV. Modulation of human platelet adenylate cyclase by prostacyclin (PGX). *Prostaglandins* 1977;**13**:377–88.
11. Weksler BB, Ley CW, Jaffe EA. Stimulation of endothelial cell prostacyclin production by thrombin, trypsin, and the ionophore A 23187. *J Clin Invest* 1978;**62**:923–30.
12. Vane JR. Nobel lecture, 8th December 1982 Adventures and excursions in bioassay: the stepping stones to prostacyclin. *Br J Pharmacol* 1983;**79**:821–38.
13. FitzGerald GA, Smith B, Pedersen AK, Brash AR. Increased prostacyclin biosynthesis in patients with severe atherosclerosis and platelet activation. *N Engl J Med* 1984;**310**:1065–8.
14. Frangos JA, Eskin SG, McIntire LV, Ives CL. Flow effects on prostacyclin production by cultured human endothelial cells. *Science* 1985;**227**:1477–9.
15. Walsh SW. Preeclampsia: an imbalance in placental prostacyclin and thromboxane production. *Am J Obstet Gynecol* 1985;**152**:335–40.
16. Walsh SW, Behr MJ, Allen NH. Placental prostacyclin production in normal and toxemic pregnancies. *Am J Obstet Gynecol* 1985;**151**:110–5.
17. de Nucci G, Thomas R, D'Orleans-Juste P, Antunes E, Walder C, Warner TD, et al. Pressor effects of circulating endothelin are limited by its removal in the pulmonary circulation and by the release of prostacyclin and endothelium-derived relaxing factor. *Proc Natl Acad Sci USA* 1988;**85**:9797–800.
18. Christman BW, McPherson CD, Newman JH, King GA, Bernard GR, Groves BM, et al. An imbalance between the excretion of thromboxane and prostacyclin metabolites in pulmonary hypertension. *N Engl J Med* 1992;**327**:70–5.

19. Boie Y, Rushmore TH, Darmon-Goodwin A, Grygorczyk R, Slipetz DM, Metters KM, et al. Cloning and expression of a cDNA for the human prostanoid IP receptor. *J Biol Chem* 1994;**269**:12173–8.
20. Katsuyama M, Sugimoto Y, Namba T, Irie A, Negishi M, Narumiya S, et al. Cloning and expression of a cDNA for the human prostacyclin receptor. *FEBS Lett* 1994;**344**:74–8.
21. Nakagawa O, Tanaka I, Usui T, Harada M, Sasaki Y, Itoh H, et al. Molecular cloning of human prostacyclin receptor cDNA and its gene expression in the cardiovascular system. *Circulation* 1994;**90**:1643–7.
22. Namba T, Oida H, Sugimoto Y, Kakizuka A, Negishi M, Ichikawa A, et al. cDNA cloning of a mouse prostacyclin receptor. Multiple signaling pathways and expression in thymic medulla. *J Biol Chem* 1994;**269**:9986–92.
23. Coleman RA, Smith WL, Narumiya S. International Union of Pharmacology classification of prostanoid receptors: properties, distribution, and structure of the receptors and their subtypes. *Pharmacol Rev* 1994;**46**:205–29.
24. Abramovitz M, Adam M, Boie Y, Grygorczyk R, Rushmore TH, Nguyen T, et al. Human prostanoid receptors: cloning and characterization. *Adv Prostagl Thromb Leukot Res* 1995;**23**:499–504.
25. Duncan AM, Anderson LL, Funk CD, Abramovitz M, Adam M. Chromosomal localization of the human prostanoid receptor gene family. *Genomics* 1995;**25**:740–2.
26. Ogawa Y, Tanaka I, Inoue M, Yoshitake Y, Isse N, Nakagawa O, et al. Structural organization and chromosomal assignment of the human prostacyclin receptor gene. *Genomics* 1995;**27**:142–8.
27. Barst RJ, Rubin LJ, Long WA, McGoon MD, Rich S, Badesch DB, et al. A comparison of continuous intravenous epoprostenol (prostacyclin) with conventional therapy for primary pulmonary hypertension. The Primary Pulmonary Hypertension Study Group. *N Engl J Med* 1996;**334**:296–302.
28. Smyth EM, Nestor PV, FitzGerald GA. Agonist-dependent phosphorylation of an epitope-tagged human prostacyclin receptor. *J Biol Chem* 1996;**271**:33698–704.
29. Forman BM, Chen J, Evans RM. Hypolipidemic drugs, polyunsaturated fatty acids, and eicosanoids are ligands for peroxisome proliferator-activated receptors alpha and delta. *Proc Natl Acad Sci USA* 1997;**94**:4312–7.
30. Murata T, Ushikubi F, Matsuoka T, Hirata M, Yamasaki A, Sugimoto Y, et al. Altered pain perception and inflammatory response in mice lacking prostacyclin receptor. *Nature* 1997;**388**:678–82.
31. Nagaya N, Uematsu M, Okano Y, Satoh T, Kyotani S, Sakamaki F, et al. Effect of orally active prostacyclin analogue on survival of outpatients with primary pulmonary hypertension. *J Am Coll Cardiol* 1999;**34**:1188–92.
32. Vinals M, Martinez-Gonzalez J, Badimon L. Regulatory effects of HDL on smooth muscle cell prostacyclin release. *Arterioscler Thromb Vasc Biol* 1999;**19**:2405–11.
33. McAdam BF, Catella-Lawson F, Mardini IA, Kapoor S, Lawson JA, FitzGerald GA. Systemic biosynthesis of prostacyclin by cyclooxygenase (COX)-2: the human pharmacology of a selective inhibitor of COX-2. *Proc Natl Acad Sci USA* 1999;**96**:272–7.
34. Catella-Lawson F, McAdam B, Morrison BW, Kapoor S, Kujubu D, Antes L, et al. Effects of specific inhibition of cyclooxygenase-2 on sodium balance, hemodynamics, and vasoactive eicosanoids. *J Pharmacol Exp Ther* 1999;**289**:735–41.
35. Hoeper MM, Schwarze M, Ehlerding S, Adler-Schuermeyer A, Spiekerkoetter E, Niedermeyer J, et al. Long-term treatment of primary pulmonary hypertension with aerosolized iloprost, a prostacyclin analogue. *N Engl J Med* 2000;**342**:1866–70.
36. Belton O, Byrne D, Kearney D, Leahy A, Fitzgerald DJ. Cyclooxygenase-1 and -2-dependent prostacyclin formation in patients with atherosclerosis. *Circulation* 2000;**102**:840–5.

37. Bombardier C, Laine L, Reicin A, Shapiro D, Burgos-Vargas R, Davis B, et al. Comparison of upper gastrointestinal toxicity of rofecoxib and naproxen in patients with rheumatoid arthritis. VIGOR Study Group. *N Engl J Med* 2000;**343**:1520–8 , 1522 p following 1528.
38. Zhang Z, Austin SC, Smyth EM. Glycosylation of the human prostacyclin receptor: role in ligand binding and signal transduction. *Mol Pharmacol* 2001;**60**:480–7.
39. Cheng Y, Austin SC, Rocca B, Koller BH, Coffman TM, Grosser T, et al. Role of prostacyclin in the cardiovascular response to thromboxane A2. *Science* 2002;**296**:539–41.
40. Stitham J, Stojanovic A, Hwa J. Impaired receptor binding and activation associated with a human prostacyclin receptor polymorphism. *J Biol Chem* 2002;**277**:15439–44.
41. Miggin SM, Lawler OA, Kinsella BT. Palmitoylation of the human prostacyclin receptor. Functional implications of palmitoylation and isoprenylation. *J Biol Chem* 2003;**278**:6947–58.
42. Stitham J, Stojanovic A, Merenick BL, O'Hara KA, Hwa J. The unique ligand-binding pocket for the human prostacyclin receptor. Site-directed mutagenesis and molecular modeling. *J Biol Chem* 2003;**278**:4250–7.
43. Ruan KH, Wu J, So SP, Jenkins LA. Evidence of the residues involved in ligand recognition in the second extracellular loop of the prostacyclin receptor characterized by high resolution 2D NMR techniques. *Arch Biochem Biophys* 2003;**418**:25–33.
44. Egan KM, Lawson JA, Fries S, Koller B, Rader DJ, Smyth EM, et al. COX-2-derived prostacyclin confers atheroprotection on female mice. *Science* 2004;**306**:1954–7.
45. Wilson SJ, Roche AM, Kostetskaia E, Smyth EM. Dimerization of the human receptors for prostacyclin and thromboxane facilitates thromboxane receptor-mediated cAMP generation. *J Biol Chem* 2004;**279**:53036–47.
46. Bresalier RS, Sandler RS, Quan H, Bolognese JA, Oxenius B, Horgan K, et al. Cardiovascular events associated with rofecoxib in a colorectal adenoma chemoprevention trial. *N Engl J Med* 2005;**352**:1092–102.
47. Nussmeier NA, Whelton AA, Brown MT, Langford RM, Hoeft A, Parlow JL, et al. Complications of the COX-2 inhibitors parecoxib and valdecoxib after cardiac surgery. *N Engl J Med* 2005;**352**:1081–91.
48. Bley KR, Bhattacharya A, Daniels DV, Gever J, Jahangir A, O'Yang C, et al. RO1138452 and RO3244794: characterization of structurally distinct, potent and selective IP (prostacyclin) receptor antagonists. *Br J Pharmacol* 2006;**147**:335–45.
49. Fetalvero KM, Shyu M, Nomikos AP, Chiu YF, Wagner RJ, Powell RJ, et al. The prostacyclin receptor induces human vascular smooth muscle cell differentiation via the protein kinase A pathway. *Am J Physiol Heart Circ Physiol* 2006;**290**:H1337–46.
50. Stitham J, Gleim SR, Douville K, Arehart E, Hwa J. Versatility and differential roles of cysteine residues in human prostacyclin receptor structure and function. *J Biol Chem* 2006;**281**:37227–336.
51. Arehart E, Stitham J, Asselbergs FW, Douville K, MacKenzie T, Fetalvero KM, et al. Acceleration of cardiovascular disease by a dysfunctional prostacyclin receptor mutation: potential implications for cyclooxygenase-2 inhibition. *Circ Res* 2008;**102**:986–93.
52. Patrignani P, Di Febbo C, Tacconelli S, Douville K, Guglielmi MD, Horvath RJ, et al. Differential association between human prostacyclin receptor polymorphisms and the development of venous thrombosis and intimal hyperplasia: a clinical biomarker study. *Pharmacogenet Genomics* 2008;**18**:611–20.
53. Fetalvero KM, Zhang P, Shyu M, Young BT, Hwa J, Young RC, et al. Prostacyclin primes pregnant human myometrium for an enhanced contractile response in parturition. *J Clin Invest* 2008;**118**:3966–79.
54. Jones RL, Qian Y, Wong HN, Chan H, Yim AP. Prostanoid action on the human pulmonary vascular system. *Clin Exp Pharmacol Physiol* 1997;**24**:969–72.
55. Dogne JM, de Leval X, Hanson J, Frederich M, Lambermont B, Ghuysen A, et al. New developments on thromboxane and prostacyclin modulators part I: thromboxane modulators. *Curr Med Chem* 2004;**11**:1223–41.

56. Tuder RM, Zaiman AL. Prostacyclin analogs as the brakes for pulmonary artery smooth muscle cell proliferation: is it sufficient to treat severe pulmonary hypertension? *Am J Respir Cell Mol Biol* 2002;**26**:171–4.
57. Olschewski H, Simonneau G, Galie N, Higenbottam T, Naeije R, Rubin LJ, et al. Inhaled iloprost for severe pulmonary hypertension. *N Engl J Med* 2002;**347**:322–9.
58. Grosser T, Fries S, Fitzgerald GA. Biological basis for the cardiovascular consequences of COX-2 inhibition: therapeutic challenges and opportunities. *J Clin Invest* 2006;**116**:4–15.
59. Solomon SD, Wittes J, Finn PV, Fowler R, Viner J, Bertagnolli MM, et al. Cardiovascular risk of Celecoxib in 6 randomized placebo-controlled trials: the cross trial safety analysis. *Circulation* 2008;**117**:2104–13.
60. Giguere V, Gallant MA, de Brum-Fernandes AJ, Parent JL. Role of extracellular cysteine residues in dimerization/oligomerization of the human prostacyclin receptor. *Eur J Pharmacol* 2004;**494**:11–22.
61. Stitham J, Gleim SR, Douville K, Arehart E, Hwa J. Versatility and differential roles of cysteine residues in human prostacyclin receptor structure and function. *J Biol Chem* 2006;**281**:37227–36.
62. Narumiya S, Sugimoto Y, Ushikubi F. Prostanoid receptors: structures, properties, and functions. *Physiol Rev* 1999;**79**:1193–226.
63. Wise HJR. *Prostacyclin and its receptors*. New York: Plenum Publishers; 2000, 310 pp.
64. Miggin SM, Kinsella BT. Investigation of the mechanisms of G protein: effector coupling by the human and mouse prostacyclin receptors. Identification of critical species-dependent differences. *J Biol Chem* 2002;**277**:27053–64.
65. Reginato MJ, Krakow SL, Bailey ST, Lazar MA. Prostaglandins promote and block adipogenesis through opposing effects on peroxisome proliferator-activated receptor gamma. *J Biol Chem* 1998;**273**:1855–8.
66. Bishop-Bailey D. Peroxisome proliferator-activated receptors in the cardiovascular system. *Br J Pharmacol* 2000;**129**:823–34.
67. Pola R, Gaetani E, Flex A, Aprahamian TR, Bosch-Marce M, Losordo DW, et al. Comparative analysis of the in vivo angiogenic properties of stable prostacyclin analogs: a possible role for peroxisome proliferator-activated receptors. *J Mol Cell Cardiol* 2004;**36**:363–70.
68. Stitham J, Arehart EJ, Gleim SR, Douville KL, Hwa J. Human prostacyclin receptor structure and function from naturally-occurring and synthetic mutations. *Prostaglandins Other Lipid Mediat* 2007;**82**:95–108.
69. Kobayashi T, Kiriyama M, Hirata T, Hirata M, Ushikubi F, Narumiya S. Identification of domains conferring ligand binding specificity to the prostanoid receptor. Studies on chimeric prostacyclin/prostaglandin D receptors. *J Biol Chem* 1997;**272**:15154–60.
70. Kobayashi T, Ushikubi F, Narumiya S. Amino acid residues conferring ligand binding properties of prostaglandin I and prostaglandin D receptors. Identification by site-directed mutagenesis. *J Biol Chem* 2000;**275**:24294–303.
71. Stitham J, Stojanovic A, Ross LA, Blount Jr AC, Hwa J. Clusters of transmembrane residues are critical for human prostacyclin receptor activation. *Biochemistry* 2004;**43**:8974–86.
72. Stitham J, Martin KA, Hwa J. The critical role of transmembrane prolines in human prostacyclin receptor activation. *Mol Pharmacol* 2002;**61**:1202–10.
73. Farrens DL, Altenbach C, Yang K, Hubbell WL, Khorana HG. Requirement of rigid-body motion of transmembrane helices for light activation of rhodopsin. *Science* 1996;**274**:768–70.
74. Capra V, Veltri A, Foglia C, Crimaldi L, Habib A, Parenti M, et al. Mutational analysis of the highly conserved ERY motif of the thromboxane A2 receptor: alternative role in G protein-coupled receptor signaling. *Mol Pharmacol* 2004;**66**:880–9.
75. O'Meara SJ, Kinsella BT. The effect of the farnesyl protein transferase inhibitor SCH66336 on isoprenylation and signalling by the prostacyclin receptor. *Biochem J* 2005;**386**:177–89.

76. Lawler OA, Miggin SM, Kinsella BT. Protein kinase A-mediated phosphorylation of serine 357 of the mouse prostacyclin receptor regulates its coupling to G(s)-, to G(i)-, and to G(q)-coupled effector signaling. *J Biol Chem* 2001;**276**:33596–607.
77. Zhang L, Huang G, Wu J, Ruan KH. A profile of the residues in the first intracellular loop critical for Gs-mediated signaling of human prostacyclin receptor characterized by an integrative approach of NMR-experiment and mutagenesis. *Biochemistry* 2005;**44**:11389–401.
78. Ruan CH, Wu J, Ruan KH. A strategy using NMR peptide structures of thromboxane A2 receptor as templates to construct ligand-recognition pocket of prostacyclin receptor. *BMC Biochem* 2005;**6**:23.
79. Owens GK, Kumar MS, Wamhoff BR. Molecular regulation of vascular smooth muscle cell differentiation in development and disease. *Physiol Rev* 2004;**84**:767–801.
80. Blank RS, Owens GK. Platelet-derived growth factor regulates actin isoform expression and growth state in cultured rat aortic smooth muscle cells. *J Cell Physiol* 1990;**142**:635–42.
81. Owens GK. Regulation of differentiation of vascular smooth muscle cells. *Physiol Rev* 1995;**75**:487–517.
82. Sobue K, Hayashi K, Nishida W. Expressional regulation of smooth muscle cell-specific genes in association with phenotypic modulation. *Mol Cell Biochem* 1999;**190**:105–18.
83. Li X, Van Putten V, Zarinetchi F, Nicks ME, Thaler S, Heasley LE, et al. Suppression of smooth-muscle alpha-actin expression by platelet-derived growth factor in vascular smooth-muscle cells involves Ras and cytosolic phospholipase A2. *Biochem J* 1997;**327**(Pt 3):709–16.
84. Holycross BJ, Blank RS, Thompson MM, Peach MJ, Owens GK. Platelet-derived growth factor-BB-induced suppression of smooth muscle cell differentiation. *Circ Res* 1992;**71**:1525–32.
85. Thyberg J, Palmberg L, Nilsson J, Ksiazek T, Sjolund M. Phenotype modulation in primary cultures of arterial smooth muscle cells. On the role of platelet-derived growth factor. *Differentiation* 1983;**25**:156–67.
86. Corjay MH, Thompson MM, Lynch KR, Owens GK. Differential effect of platelet-derived growth factor- versus serum-induced growth on smooth muscle alpha-actin and nonmuscle beta-actin mRNA expression in cultured rat aortic smooth muscle cells. *J Biol Chem* 1989;**264**:10501–6.
87. Owens GK, Thompson MM. Developmental changes in isoactin expression in rat aortic smooth muscle cells in vivo. Relationship between growth and cytodifferentiation. *J Biol Chem* 1986;**261**:13373–80.
88. Cook CL, Weiser MC, Schwartz PE, Jones CL, Majack RA. Developmentally timed expression of an embryonic growth phenotype in vascular smooth muscle cells. *Circ Res* 1994;**74**:189–96.
89. O'Brien ER, Alpers CE, Stewart DK, Ferguson M, Tran N, Gordon D, et al. Proliferation in primary and restenotic coronary atherectomy tissue. Implications for antiproliferative therapy. *Circ Res* 1993;**73**:223–31.
90. Wilcox JN. Analysis of local gene expression in human atherosclerotic plaques. *J Vasc Surg* 1992;**15**:913–6.
91. Fillinger MF, Sampson LN, Cronenwett JL, Powell RJ, Wagner RJ. Coculture of endothelial cells and smooth muscle cells in bilayer and conditioned media models. *J Surg Res* 1997;**67**:169–78.
92. Brown DJ, Rzucidlo EM, Merenick BL, Wagner RJ, Martin KA, Powell RJ. Endothelial cell activation of the smooth muscle cell phosphoinositide 3-kinase/Akt pathway promotes differentiation. *J Vasc Surg* 2005;**41**:509–16.
93. Vane JR, Anggard EE, Botting RM. Regulatory functions of the vascular endothelium. *N Engl J Med* 1990;**323**:27–36.
94. Gryglewski RJ, Botting RM, Vane JR. Mediators produced by the endothelial cell. *Hypertension* 1988;**12**:530–48.

95. Boger RH, Bode-Boger SM, Thiele W, Junker W, Alexander K, Frolich JC. Biochemical evidence for impaired nitric oxide synthesis in patients with peripheral arterial occlusive disease. *Circulation* 1997;**95**:2068–74.
96. Feron O, Dessy C, Moniotte S, Desager JP, Balligand JL. Hypercholesterolemia decreases nitric oxide production by promoting the interaction of caveolin and endothelial nitric oxide synthase. *J Clin Invest* 1999;**103**:897–905.
97. Munzel T, Daiber A, Ullrich V, Mulsch A. Vascular consequences of endothelial nitric oxide synthase uncoupling for the activity and expression of the soluble guanylyl cyclase and the cGMP-dependent protein kinase. *Arterioscler Thromb Vasc Biol* 2005;**25**:1551–7.
98. Lopez JA, Armstrong ML, Piegors DJ, Heistad DD. Vascular responses to endothelin-1 in atherosclerotic primates. *Arteriosclerosis* 1990;**10**:1113–8.
99. Levin ER. Endothelins. *N Engl J Med* 1995;**333**:356–63.
100. Simonson MS, Dunn MJ. Cellular signaling by peptides of the endothelin gene family. *FASEB J* 1990;**4**:2989–3000.
101. McIntyre TM, Zimmerman GA, Satoh K, Prescott SM. Cultured endothelial cells synthesize both platelet-activating factor and prostacyclin in response to histamine, bradykinin, and adenosine triphosphate. *J Clin Invest* 1985;**76**:271–80.
102. Zardi EM, Zardi DM, Cacciapaglia F, Dobrina A, Amoroso A, Picardi A, et al. Endothelial dysfunction and activation as an expression of disease: role of prostacyclin analogs. *Int Immunopharmacol* 2005;**5**:437–59.
103. Cassano S, Di Lieto A, Cerillo R, Avvedimento EV. Membrane-bound cAMP-dependent protein kinase controls cAMP-induced differentiation in PC12 cells. *J Biol Chem* 1999;**274**:32574–9.
104. Stork PJ, Schmitt JM. Crosstalk between cAMP and MAP kinase signaling in the regulation of cell proliferation. *Trends Cell Biol* 2002;**12**:258–66.
105. Murray R, Shipp E, FitzGerald GA. Prostaglandin endoperoxide/thromboxane A2 receptor desensitization. Cross-talk with adenylate cyclase in human platelets. *J Biol Chem* 1990;**265**:21670–5.
106. Hara S, Morishita R, Tone Y, Yokoyama C, Inoue H, Kaneda Y, et al. Overexpression of prostacyclin synthase inhibits growth of vascular smooth muscle cells. *Biochem Biophys Res Commun* 1995;**216**:862–7.
107. Todaka T, Yokoyama C, Yanamoto H, Hashimoto N, Nagata I, Tsukahara T, et al. Gene transfer of human prostacyclin synthase prevents neointimal formation after carotid balloon injury in rats. *Stroke* 1999;**30**:419–26.
108. Kothapalli D, Stewart SA, Smyth EM, Azonobi I, Pure E, Assoian RK. Prostacylin receptor activation inhibits proliferation of aortic smooth muscle cells by regulating cAMP response element-binding protein- and pocket protein-dependent cyclin a gene expression. *Mol Pharmacol* 2003;**64**:249–58.
109. Stewart SA, Kothapalli D, Yung Y, Assoian RK. Antimitogenesis linked to regulation of Skp2 gene expression. *J Biol Chem* 2004;**279**:29109–13.
110. DeSalle LM, Pagano M. Regulation of the G1 to S transition by the ubiquitin pathway. *FEBS Lett* 2001;**490**:179–89.
111. Weinberg RA. The retinoblastoma protein and cell cycle control. *Cell* 1995;**81**:323–30.
112. Ii M, Hoshiga M, Fukui R, Negoro N, Nakakoji T, Nishiguchi F, et al. Beraprost sodium regulates cell cycle in vascular smooth muscle cells through cAMP signaling by preventing down-regulation of p27(Kip1). *Cardiovasc Res* 2001;**52**:500–8.
113. Sheaff RJ, Groudine M, Gordon M, Roberts JM, Clurman BE. Cyclin E-CDK2 is a regulator of p27Kip1. *Genes Dev* 1997;**11**:1464–78.
114. Nakayama K, Nagahama H, Minamishima YA, Matsumoto M, Nakamichi I, Kitagawa K, et al. Targeted disruption of Skp2 results in accumulation of cyclin E and p27(Kip1), polyploidy and centrosome overduplication. *EMBO J* 2000;**19**:2069–81.

115. Ganoth D, Bornstein G, Ko TK, Larsen B, Tyers M, Pagano M, et al. The cell-cycle regulatory protein Cks1 is required for SCF(Skp2)-mediated ubiquitinylation of p27. *Nat Cell Biol* 2001;**3**:321–4.
116. Spruck C, Strohmaier H, Watson M, Smith AP, Ryan A, Krek TW, et al. A CDK-independent function of mammalian Cks1: targeting of SCF(Skp2) to the CDK inhibitor p27Kip1. *Mol Cell* 2001;**7**:639–50.
117. Braun-Dullaeus RC, Mann MJ, Seay U, Zhang L, von Der Leyen HE, Morris RE, et al. Cell cycle protein expression in vascular smooth muscle cells *in vitro* and *in vivo* is regulated through phosphatidylinositol 3-kinase and mammalian target of rapamycin. *Arterioscler Thromb Vasc Biol* 2001;**21**:1152–8.
118. Chen D, Krasinski K, Sylvester A, Chen J, Nisen PD, Andres V. Downregulation of cyclin-dependent kinase 2 activity and cyclin A promoter activity in vascular smooth muscle cells by p27(KIP1), an inhibitor of neointima formation in the rat carotid artery. *J Clin Invest* 1997;**99**:2334–41.
119. Diez-Juan A, Castro C, Edo MD, Andres V. Role of the growth suppressor p27Kip1 during vascular remodeling. *Curr Vasc Pharmacol* 2003;**1**:99–106.
120. Blindt R, Bosserhoff AK, Vom Dahl J, Hanrath P, Schror K, Hohlfeld T, et al. Activation of IP and EP(3) receptors alters cAMP-dependent cell migration. *Eur J Pharmacol* 2002;**444**:31–7.
121. Bulin C, Albrecht U, Bode JG, Weber AA, Schror K, Levkau B, et al. Differential effects of vasodilatory prostaglandins on focal adhesions, cytoskeletal architecture, and migration in human aortic smooth muscle cells. *Arterioscler Thromb Vasc Biol* 2005;**25**:84–9.
122. Hansson GK, Libby P. The immune response in atherosclerosis: a double-edged sword. *Nat Rev Immunol* 2006;**6**:508–19.
123. Baker CS, Hall RJ, Evans TJ, Pomerance A, Maclouf J, Creminon C, et al. Cyclooxygenase-2 is widely expressed in atherosclerotic lesions affecting native and transplanted human coronary arteries and colocalizes with inducible nitric oxide synthase and nitrotyrosine particularly in macrophages. *Arterioscler Thromb Vasc Biol* 1999;**19**:646–55.
124. Schonbeck U, Sukhova GK, Graber P, Coulter S, Libby P. Augmented expression of cyclooxygenase-2 in human atherosclerotic lesions. *Am J Pathol* 1999;**155**:1281–91.
125. Stemme V, Swedenborg J, Claesson H, Hansson GK. Expression of cyclo-oxygenase-2 in human atherosclerotic carotid arteries. *Eur J Vasc Endovasc Surg* 2000;**20**:146–52.
126. Belton O, Byrne D, Kearney D, Leahy A, Fitzgerald DJ. Cyclooxygenase-1 and -2-dependent prostacyclin formation in patients with atherosclerosis. *Circulation* 2000;**102**:840–5.
127. Cipollone F, Prontera C, Pini B, Marini M, Fazia M, De Cesare D, et al. Overexpression of functionally coupled cyclooxygenase-2 and prostaglandin E synthase in symptomatic atherosclerotic plaques as a basis of prostaglandin E(2)-dependent plaque instability. *Circulation* 2001;**104**:921–7.
128. FitzGerald GA, Pedersen AK, Patrono C. Analysis of prostacyclin and thromboxane biosynthesis in cardiovascular disease. *Circulation* 1983;**67**:1174–7.
129. Clarke RJ, Mayo G, Price P, FitzGerald GA. Suppression of thromboxane A2 but not of systemic prostacyclin by controlled-release aspirin. *N Engl J Med* 1991;**325**:1137–41.
130. Cipollone F, Fazia M, Iezzi A, Ciabattoni G, Pini B, Cuccurullo C, et al. Balance between PGD synthase and PGE synthase is a major determinant of atherosclerotic plaque instability in humans. *Arterioscler Thromb Vasc Biol* 2004;**24**:1259–65.
131. Della Bella S, Molteni M, Mocellin C, Fumagalli S, Bonara P, Scorza R. Novel mode of action of iloprost: *in vitro* down-regulation of endothelial cell adhesion molecules. *Prostaglandins Other Lipid Mediat* 2001;**65**:73–83.
132. Pomerantz KB, Fleisher LN, Tall AR, Cannon PJ. Enrichment of endothelial cell arachidonate by lipid transfer from high density lipoproteins: relationship to prostaglandin I2 synthesis. *J Lipid Res* 1985;**26**:1269–76.

133. Vinals M, Martinez-Gonzalez J, Badimon JJ, Badimon L. HDL-induced prostacyclin release in smooth muscle cells is dependent on cyclooxygenase-2 (Cox-2). *Arterioscler Thromb Vasc Biol* 1997;**17**:3481–8.
134. Thiemermann C. Biosynthesis and interaction of endothelium-derived vasoactive mediators. *Eicosanoids* 1991;**4**:187–202.
135. Weksler BB, Hajjar DP, Eldor A, Falcone DJ, Tack-Goldman K, Minick CR. Interactions between prostacyclin metabolism and cholesteryl ester metabolism in the vascular wall. *Adv Prostaglandin Thromb Leukot Res* 1983;**11**:463–7.
136. Hajjar DP, Weksler BB. Metabolic activity of cholesteryl esters in aortic smooth muscle cells is altered by prostaglandins I2 and E2. *J Lipid Res* 1983;**24**:1176–85.
137. Hajjar DP, Marcus AJ, Etingin OR. Platelet-neutrophil-smooth muscle cell interactions: lipoxygenase-derived mono- and dihydroxy acids activate cholesteryl ester hydrolysis by the cyclic AMP dependent protein kinase cascade. *Biochemistry* 1989;**28**:8885–91.
138. Bunting S, Moncada S, Vane JR. The prostacyclin–thromboxane A2 balance: pathophysiological and therapeutic implications. *Br Med Bull* 1983;**39**:271–6.
139. Davies P, Bailey PJ, Goldenberg MM, Ford-Hutchinson AW. The role of arachidonic acid oxygenation products in pain and inflammation. *Annu Rev Immunol* 1984;**2**:335–57.
140. Williams TJ. Prostaglandin E2, prostaglandin I2 and the vascular changes of inflammation. *Br J Pharmacol* 1979;**65**:517–24.
141. Bley KR, Hunter JC, Eglen RM, Smith JA. The role of IP prostanoid receptors in inflammatory pain. *Trends Pharmacol Sci* 1998;**19**:141–7.
142. Smith WL. The eicosanoids and their biochemical mechanisms of action. *Biochem J* 1989;**259**:315–24.
143. Berkenkopf JW, Weichman BM. Production of prostacyclin in mice following intraperitoneal injection of acetic acid, phenylbenzoquinone and zymosan: its role in the writhing response. *Prostaglandins* 1988;**36**:693–709.
144. Doherty NS, Beaver TH, Chan KY, Coutant JE, Westrich GL. The role of prostaglandins in the nociceptive response induced by intraperitoneal injection of zymosan in mice. *Br J Pharmacol* 1987;**91**:39–47.
145. Akarsu ES, Palaoglu O, Ayhan IH. Iloprost-induced writhing in mice and its suppression by morphine. *Methods Find Exp Clin Pharmacol* 1989;**11**:273–5.
146. McQueen DS, Iggo A, Birrell GJ, Grubb BD. Effects of paracetamol and aspirin on neural activity of joint mechanonociceptors in adjuvant arthritis. *Br J Pharmacol* 1991;**104**:178–82.
147. Ferreira SH, Nakamura M. I—Prostaglandin hyperalgesia, a cAMP/Ca2+ dependent process. *Prostaglandins* 1979;**18**:179–90.
148. Malmberg AB, Brandon EP, Idzerda RL, Liu H, McKnight GS, Basbaum AI. Diminished inflammation and nociceptive pain with preservation of neuropathic pain in mice with a targeted mutation of the type I regulatory subunit of cAMP-dependent protein kinase. *J Neurosci* 1997;**17**:7462–70.
149. Sugita S, Baxter DA, Byrne JH. Modulation of a cAMP/protein kinase A cascade by protein kinase C in sensory neurons of Aplysia. *J Neurosci* 1997;**17**:7237–44.
150. Schepelmann K, Messlinger K, Schmidt RF. The effects of phorbol ester on slowly conducting afferents of the cat's knee joint. *Exp Brain Res* 1993;**92**:391–8.
151. Malmberg AB, Chen C, Tonegawa S, Basbaum AI. Preserved acute pain and reduced neuropathic pain in mice lacking PKCgamma. *Science* 1997;**278**:279–83.
152. Hingtgen CM, Waite KJ, Vasko MR. Prostaglandins facilitate peptide release from rat sensory neurons by activating the adenosine $3'$, $5'$-cyclic monophosphate transduction cascade. *J Neurosci* 1995;**15**:5411–9.
153. Nicol GD, Vasko MR, Evans AR. Prostaglandins suppress an outward potassium current in embryonic rat sensory neurons. *J Neurophysiol* 1997;**77**:167–76.

154. England S, Bevan S, Docherty RJ. PGE2 modulates the tetrodotoxin-resistant sodium current in neonatal rat dorsal root ganglion neurones via the cyclic AMP-protein kinase A cascade. *J Physiol* 1996;**495**(Pt 2):429–40.
155. Undem BJ, Weinrich D. Electrophysiological properties and chemosensitivity of guinea pig nodose ganglion neurons *in vitro*. *J Auton Nerv Syst* 1993;**44**:17–33.
156. Pitchford S, Levine JD. Prostaglandins sensitize nociceptors in cell culture. *Neurosci Lett* 1991;**132**:105–8.
157. Xiao CY, Hara A, Yuhki Ki K, Fujino T, Ma H, Okada Y, et al. Roles of prostaglandin i(2) and thromboxane a(2) in cardiac ischemia-reperfusion injury: a study using mice lacking their respective receptors. *Circulation* 2001;**104**:2210–5.
158. Challis JR, Sloboda DM, Alfaidy N, Lye SJ, Gibb W, Patel FA, et al. Prostaglandins and mechanisms of preterm birth. *Reproduction* 2002;**124**:1–17.
159. Hertelendy F, Zakar T. Prostaglandins and the myometrium and cervix. *Prostaglandins Leukot Essent Fatty Acids* 2004;**70**:207–22.
160. Bennett PR, Elder MG, Myatt L. The effects of lipoxygenase metabolites of arachidonic acid on human myometrial contractility. *Prostaglandins* 1987;**33**:837–44.
161. Ritchie DM, Hahn DW, McGuire JL. Smooth muscle contraction as a model to study the mediator role of endogenous lipoxygenase products of arachidonic acid. *Life Sci* 1984;**34**:509–13.
162. Wikland M, Lindblom B, Wiqvist N. Myometrial response to prostaglandins during labor. *Gynecol Obstet Invest* 1984;**17**:131–8.
163. Wiqvist N, Lindblom B, Wikland M, Wilhelmsson L. Prostaglandins and uterine contractility. *Acta Obstet Gynecol Scand Suppl* 1983;**113**:23–9.
164. Abel MH, Kelly RW. Differential production of prostaglandins within the human uterus. *Prostaglandins* 1979;**18**:821–8.
165. Bamford DS, Jogee M, Williams KI. Prostacyclin formation by the pregnant human myometrium. *Br J Obstet Gynaecol* 1980;**87**:215–8.
166. Omini C, Folco GC, Pasargiklian R, Fano M, Berti F. Prostacyclin (PGI2) in pregnant human uterus. *Prostaglandins* 1979;**17**:113–20.
167. Korita D, Itoh H, Sagawa N, Yura S, Yoshida M, Kakui K, et al. 17beta-estradiol up-regulates prostacyclin production in cultured human uterine myometrial cells via augmentation of both cyclooxygenase-1 and prostacyclin synthase expression. *J Soc Gynecol Investig* 2004;**11**:457–64.
168. Ylikorkala O, Makarainen L, Viinikka L. Prostacyclin production increases during human parturition. *Br J Obstet Gynaecol* 1981;**88**:513–6.
169. Ylikorkala O, Paatero H, Suhonen L, Viinikka L. Vaginal and abdominal delivery increases maternal urinary 6-keto-prostaglandin F1 alpha excretion. *Br J Obstet Gynaecol* 1986;**93**:950–4.
170. Omini C, Pasargiklian R, Folco GC, Fano M, Berti F. Pharmacological activity of PGI2 and its metabolite 6-oxo-PGF1alpha on human uterus and fallopian tubes. *Prostaglandins* 1978;**15**:1045–54.
171. Korita D, Sagawa N, Itoh H, Yura S, Yoshida M, Kakui K, et al. Cyclic mechanical stretch augments prostacyclin production in cultured human uterine myometrial cells from pregnant women: possible involvement of up-regulation of prostacyclin synthase expression. *J Clin Endocrinol Metab* 2002;**87**:5209–19.
172. Armstrong JM, Lattimer N, Moncada S, Vane JR. Comparison of the vasodepressor effects of prostacyclin and 6-oxo-prostaglandin F1alpha with those of prostaglandin E2 in rats and rabbits. *Br J Pharmacol* 1978;**62**:125–30.
173. Gryglewski RJ, Bunting S, Moncada S, Flower RJ, Vane JR. Arterial walls are protected against deposition of platelet thrombi by a substance (prostaglandin X) which they make from prostaglandin endoperoxides. *Prostaglandins* 1976;**12**:685–713.

174. Omini C, Moncada S, Vane JR. The effects of prostacyclin (PGI2) on tissues which detect prostaglandins (PG'S). *Prostaglandins* 1977;**14**:625–32.
175. Williams KI, El-Tahir KE, Marcinkiewicz E. Dual actions of prostacyclin (PGI2) on the rat pregnant uterus. *Prostaglandins* 1979;**17**:667–72.
176. Senior J, Marshall K, Sangha R, Clayton JK. In vitro characterization of prostanoid receptors on human myometrium at term pregnancy. *Br J Pharmacol* 1993;**108**:501–6.
177. Wikland M, Lindblom B, Hammarstrom S, Wiqvist N. The effect of prostaglandin I on the contractility of the term pregnant human myometrium. *Prostaglandins* 1983;**26**:905–16.
178. Humbert M, Sitbon O, Simonneau G. Treatment of pulmonary arterial hypertension. *N Engl J Med* 2004;**351**:1425–36.
179. Rich S, Dantzker DR, Ayres SM, Bergofsky EH, Brundage BH, Detre KM, et al. Primary pulmonary hypertension. A national prospective study. *Ann Intern Med* 1987;**107**:216–23.
180. Rubin LJ. Primary pulmonary hypertension. *N Engl J Med* 1997;**336**:111–7.
181. Rabinovitch M. Pulmonary hypertension: updating a mysterious disease. *Cardiovasc Res* 1997;**34**:268–72.
182. Botney M. Epoprostenol (prostacyclin) therapy in primary pulmonary hypertension. *N Engl J Med* 1998;**338**:1773–4.
183. Sitbon O, Humbert M, Nunes H, Parent F, Garcia G, Herve P, et al. Long-term intravenous epoprostenol infusion in primary pulmonary hypertension: prognostic factors and survival. *J Am Coll Cardiol* 2002;**40**:780–8.
184. Simonneau G, Barst RJ, Galie N, Naeije R, Rich S, Bourge RC, et al. Continuous subcutaneous infusion of treprostinil, a prostacyclin analogue, in patients with pulmonary arterial hypertension: a double-blind, randomized, placebo-controlled trial. *Am J Respir Crit Care Med* 2002;**165**:800–4.
185. Galie N, Humbert M, Vachiery JL, Vizza CD, Kneussl M, Manes A, et al. Effects of beraprost sodium, an oral prostacyclin analogue, in patients with pulmonary arterial hypertension: a randomized, double-blind, placebo-controlled trial. *J Am Coll Cardiol* 2002;**39**:1496–502.
186. Saito S, Iida A, Sekine A, Kawauchi S, Higuchi S, Ogawa C, et al. Catalog of 178 variations in the Japanese population among eight human genes encoding G protein-coupled receptors (GPCRs). *J Hum Genet* 2003;**48**:461–8.
187. Stitham J, Arehart EJ, Gleim S, Douville K, MacKenzie T, Hwa J. Arginine (CGC) codon targeting in the human prostacyclin receptor gene (PTGIR) and G-protein coupled receptors (GPCR). *Gene* 2007;**396**:180–7.
188. Liggett SB, Mialet-Perez J, Thaneemit-Chen S, Weber SA, Greene SM, Hodne D, et al. A polymorphism within a conserved beta(1)-adrenergic receptor motif alters cardiac function and beta-blocker response in human heart failure. *Proc Natl Acad Sci USA* 2006;**103**:11288–93.
189. Liggett SB, Wagoner LE, Craft LL, Hornung RW, Hoit BD, McIntosh TC, et al. The Ile164 beta2-adrenergic receptor polymorphism adversely affects the outcome of congestive heart failure. *J Clin Invest* 1998;**102**:1534–9.
190. Taylor DR, Drazen JM, Herbison GP, Yandava CN, Hancox RJ, Town GI. Asthma exacerbations during long term beta agonist use: influence of beta(2) adrenoceptor polymorphism. *Thorax* 2000;**55**:762–7.
191. Warner TD, Giuliano F, Vojnovic I, Bukasa A, Mitchell JA, Vane JR. Nonsteroid drug selectivities for cyclo-oxygenase-1 rather than cyclo-oxygenase-2 are associated with human gastrointestinal toxicity: a full in vitro analysis. *Proc Natl Acad Sci USA* 1999;**96**:7563–8.
192. Barnett J, Chow J, Ives D, Chiou M, Mackenzie R, Osen E, et al. Purification, characterization and selective inhibition of human prostaglandin G/H synthase 1 and 2 expressed in the baculovirus system. *Biochim Biophys Acta* 1994;**1209**:130–9.
193. Laneuville O, Breuer DK, Dewitt DL, Hla T, Funk CD, Smith WL. Differential inhibition of human prostaglandin endoperoxide H synthases-1 and -2 by nonsteroidal anti-inflammatory drugs. *J Pharmacol Exp Ther* 1994;**271**:927–34.

Index

A

Actin fiber, reorganization, 9
Adenylate cyclase, 17, 48, 150
Adenylyl cyclase (AC), 17–18, 48, 138
ADH. *See* Autosomal dominant hypocalcemia
Adhesion proteins, 2, 7, 9
A-kinase anchoring proteins, 17
Amenorrhea, 105, 107, 118
β-Amyloid peptides, 35
AQP2 mutations, 23
Aquaporin-2 gene, 16
Arginine vasopressin (AVP), 15–16
  to increase water permeability, 48
  multiple actions of, 17
  stimulated cAMP production, dose-dependent, 22
  in water reabsorptive capacity of kidney, 20
Aromatic amino acids, 47
β-Arrestins, 26
Atherothrombosis, 133, 145, 152–154, 157
Autoimmune disorders, 66
Autoimmune hypothyroidism, 66
Autoimmune polyendocrine syndrome type 1 (APS1), 66
Autosomal dominant hypocalcemia, 32, 52–53
Autosomal recessive
  disorder, 16
  NDI, 23
  syndrome, 5
AVP-elicited osmotic water permeability, 20
AVP receptor 2 *(AVPR2)* gene, 15
AVPR2 missense mutations. *See* Missense mutations
AVPR2 mutations, 19–21, 25–26
AVP type 2 receptors (AVPR2), 17

B

Bartter's syndrome subtype V, 53–54
Bartter syndrome, 16, 23
Barttin, 54
Bilateral frontoparietal polymicrogyria (BFPP), 4–7, 9
Bone mineral density (BMD), 70, 73
Burn injury, 73

C

Cadherin, 2
Cadmium, 59
Ca^{2+}-induced IP3 production, 44
Ca^{2+}-influx, 45
Calbindin D, 50
Calcilytics, 36
Calcimimetic allosteric modulators, 43
Calcimimetic NPS R-568, 66
Calcimimetics, 36
Calcium homeostasis, 32–33
Calhex 231, 36, 43
Calindol, 36, 43
Calpains, 49
cAMP. *See* Cyclic adenosine monophosphate
Cancer metastasis, 1, 8.
  See also GPR56 protein
Cardiovascular disease, 138, 148, 152–153, 155
CASR. *See* Cell-surface calcium sensing receptor
CASR gene, 52
CASR-HEK cells, 37
Caveolin-1, 48
CCAAT box, 34
CD9/CD81/GPR56 complexes, 9
Cell-surface calcium sensing receptor
  allosteric modifiers in clinic, 73
  allosteric modulators of, 36
  altered expression of, 71–73

Cell-surface calcium sensing receptor (cont.)
  autoantibodies and, 66
  and diseases, 33
  disorders associated with, 50–55
  downregulation and, 45
  family C GPCR, 34
  functional characterization of
    asparagine-linked glycosylation, 37
    COOH-terminal tail, 44–45
    cysteine-rich domain, 40
    dimerization, 39
    extracellular loops, 42
    intracellular loops, 44
    ligand-binding sites in, 39–40
    peptide linker, 41–42
    transmembrane domain and, 42–44
    venus-flytrap domain, 37–38
  mutations
    activating, 59, 65–69
    inactivating, 54–65
  and parathyroid, 48–49
  polymorphisms, 66, 69–70
  and renal tubule, 50
  trafficking, 45
Channel-like 11-span transmembrane protein, 3
CIC-Kb function, 54
Cinacalcet, 36
Cinacalcet HCl, 73
Cirrhosis, 23
c-Jun N-kinases (JNK), 47
Class B GPCRs, 2
Cobblestone cortex-causative genes, 5
Cobblestone-like cortical malformation, 5
Cobblestone lissencephaly, 5
Congenital NDI, 15
Cortical thick ascending limb, 50, 72
COS7 cells, 45
Coxibs, 152–153
COX-2 inhibition, 155
COX-1/2 inhibitors, 153, 155
C346S mutation, 5
CTAL. See Cortical thick ascending limb
C-terminal isoprenylation motif (CSLC), 142
CVD. See Cardiovascular disease
Cyclic adenosine monophosphate, 46, 101, 117, 138
  analog, 146–148
  binding domain, 17

CASR inhibition, of hormone-stimulated, 150
  cicaprost-mediated, 142
  dependent activation of PKA, 142
  elevating agents, 147
  increase concentration, via prostacyclin, 150
  inhibiting ROMK and NKCC2 activities, 54
  mediated incorporation of water channels, 17–18
  in parathyroid cells, 48
  regulated guanine nucleotide exchange factor, 17
  signaling in platelets, 155
  and stimulation of steroidogenesis, 99
  in WT FSHR., 23
Cysteine-rich region
  from amino acids 542 to 598, role in, 38
  at carboxyl end of ECD, 117
  in CASR, 40
  conformational change to, 35
  mutations occurs within, 58
  and a peptide linker, 34
Cystinosis, 16
Cystinuria, 23

D

Deafness, 16
Dehydrated and malnourished infant, with NDI, 21
D578G mutation, 101
Diabetes insipidus, 16
Diabetes mellitus, 16
DIDMOAD syndrome, 16
Dimerization, 57. See also CASR; FSHR
  constitutive, 117
  expression of wt receptor due to, 106
  of LHCGR, 103
  status, 37
  through ECD, 39
Disorders, associated with CASR, 55
Distal convoluted tubule (DCT), 50

E

E837A mutant, 43
ECD. See Extracellular domain
ECM. See Extracellular matrix

INDEX

Endoplasmic reticulum, 6, 23–24, 34, 105–106, 119
Endothelin-1 (ET-1), 146
Epithelial sodium channel (ENaC), 23
E-prostanoid-3 receptors, 20
ER. *See* Endoplasmic reticulum
ERK1/2 activation, 66
ERK1/2 phosphorylation, 46
Estrogen, 99
E242X mutant mice, 24
Extracellular domain, 34, 37–38, 42, 59, 99–100, 117–118, 122, 124
Extracellular loop 2 (ECL2), 42, 58
Extracellular matrix, 5, 7, 145

F

F-actin, 17
Familial hypocalciuric hypercalcemia, 32, 50–51, 57
  molecular genetics of, 52
Familial isolated hypoparathyroidism (FIH), 52, 59, 71
Farnesyl protein transferase enzyme (FTase), 142
FAVLM sequence, 3
F180C mutation, 58
FHH. *See* Familial hypocalciuric hypercalcemia
FIHP. *See* Familial isolated hyperparathyroidism
Filamin-A, 48
Fluorescence *in situ* hybridization (FISH), 52
Follicle stimulating hormone receptor (FSHR)
  activation of, 123
  coupled to stimulatory heterotrimeric G protein, 117
  dimerization, 117
  endogenous activation, 117
  in females, 116
  gene, 117
  misfolded inactivating, 119
  mutants, 121–124
    constitutively active, 123–125
  mutations, 118, 122–123
  negatively charged residue, need of, 125
  recombinant, 117
  residual activity, 120

  in Sertoli cells, 118
  signaling, 121
FSH stimulation, 120
Fukuyama-type congenital muscular dystrophy (FCMD), 5

G

Gamma-aminobutyric acid ($GABA_B$) receptors, 34
Gαq-mediated activation, 142
Geneticin, 24
Genetic variants
  found from sequencing, 156
  and influence on disease, 154–155
G_i isoforms, 48
G_i proteins, 35
Gitelman syndrome, 23
Glial cells missing (GCM) elements, 34
Glial cells missing-2 (GCM2) gene, 52–53
Glycoprotein-Hormone Receptors Information Systems database, 97
Glycosylation, 37, 42, 44, 116, 138. *See also* N-linked glycosylation
Gonadotropins, 118
GPCR kinases (GRKs), 142
GPCR Natural Variants Database, 98
GPCR proteolytic site
  mediated protein cleavage, 3
  mediated proteolysis, 3
  motif, 2–3
GPCR. *See* G protein coupled receptor
GPR56 gene, 1, 4, 9
Gpr56 knockout mice, 5
G protein, 46, 99, 103, 120, 142, 144, 157
G protein coupled receptor, 1, 10, 18, 33, 42, 46, 120, 138
GPR56 protein
  biochemical properties of, 3 (*see also* N-linked glycosylation)
  in brain development
    pial basement membrane (BM), 7–8
  and brain malformation, 4–7
  and cancer, 8–9
    oncogenic property, 8–9
    tumor progression, suppressor of, 8

GPR56 protein (cont.)
  mutations identified, 5–7
    associated with BFPP, 6
    missense mutations, 6
    protein domain and mutations, 7
  signaling, 9
GPS. See GPCR proteolytic site
G_q-coupled pathway, 48
$G_{q/11}$ protein, 35, 99
G_s protein, 35
Guanine-nucleotide (G)-protein-coupled receptor, 17

H

Haploview program, 66
hCG. See Human chorionic gonadotropin
hCG signaling, 105
HEK-expressed mIP, 143
Hepatocarcinomas, 23
Hereditary neurohypophyseal diabetes insipidus, 16
hFSHR (hFSHRHB) complex, 100
hFSHR. See Human follitropin receptor
hIP receptor
  cytoplasmic region of, 142
  farnesylated at C-terminal, 142
  genetic variants of, 154
  inflammation and pain, role in, 150–151
  NMR-based structures of, 144–145
  in parturition, role in, 151–152
  sequence not found in, 143
  signal transduction, 143
  TM proline residues within, 141
Hormone-occupied FSHR$_{HB}$, 117
HUGO Gene Nomenclature Committee, 16
Human *AVPR2* gene, 17
Human CASR, 34–35. See also Cell-surface calcium sensing receptor
Human chorionic gonadotropin, 98, 116
Human follitropin receptor, 100, 102, 117, 124
Human *GPR56* gene, 2
Human lutropin receptor, 97
  activating mutations, naturally occurring, 101
  gene encoding, 100
  and human physiology, 98–99

LHCGR gene, 99–101
LHCGR mutation
  activating mutations, 101–103
  inactivating, 103–108
  mutants, 98, 106
  protein, 99–100
  variants, 107
Human prostacyclin receptor gene (PTGIR), 138
Human thyrotropin receptor (hTSHR), 100
Hydroperoxyeicosatetranoic acid (PHETE), 49
15-Hydroxyeicosatetranoic acid (HETE), 49
Hypercalciuria, 53, 72
  predictor of, 70
Hypergonadotropic hypogonadism, 118
Hyperparathyroidism, 71, 73
Hyperphosphatemia, 53
Hyperreninemia, 53
Hypocalcemia, 53–54
Hypocalciuric hypercalcemia (AHH), 66
Hypokalemic alkalosis, 53
Hyponatremia, 24
Hypoparathyroidism, 53

I

Idiopathic epilepsy, 66
Immunoglobulin, 2
Inactivating mutations
  of *CASR* gene, 33, 60–65
  FSHR mutations, 118–121
  in ICD, 59
  of LHCGR, 103–108
  missense mutations, 57
Inner medullary collecting duct (IMCD), 50
Inositol trisphosphate (IP3), 44
Integrins, 8
Interleukin-1β (IL-1β), 73
Interleukin-6 (IL-6), 73
Intracellular cAMP receptors, 17

J

Juvenile hepatitis, 23

INDEX

L

L-α-amino acids, 34
Laminin, 5, 7
Laminin–integrin interactions, 8
LCH. *See* Leydig cell hypoplasia
Lectin, 2
Leucine-rich repeats (LRRs), 99
Leydig cell hypoplasia, 105, 107–108
Leydig cells, 98, 104
LHCGR. *See* Human lutropin receptor; LH/chorionic gonadotropin (CG) receptor
LH/chorionic gonadotropin (CG) receptor, 116
   constitutive activation in, 123–124
   homologous mutations, 123
   mutants, 119, 122, 124
   mutations in TSHR and, 121, 123
   Org41841 activating, 121
LH/FSH ratio, 105
LH signaling, 105
Ligand activation, of CASR dimer, 35
Lissencephaly, 5
Lithium, 16
Long N-terminal class B 7-transmembrane proteins (LNB-7TM), 2
Luteinizing hormone (LH), 116
ΔL608,V609 mutant, 105

M

MacSpartan Pro software, 143–144
MAPK signaling pathway, 57
Medullary thick ascending limb (MTAL), 50
Metabotropic receptor (class C), 2
Microcephaly, 5
mIP receptor, 143
Missense mutations, 7, 21
   from amino acids 250–530, 58
   AVPR2, 23
   inactivating, 57
   LHCGR, 105
   R898G, 65
   in tip of GPR56^N, 6
Mitogen-activated protein kinase (MAPK), 46
mouse prostaglandin D (mDP) receptors, 139
mouse prostaglandin I (mIP) receptors, 139
Muscle–eye–brain (MEB) disease, 5
Mutant V_2 receptors, 22
Mutations CASR, 56
Mutations in human *GPR56* gene, 1

N

NDI. *See* Nephrogenic diabetes insipidus
Neonatal severe hyperparathyroidism, 33, 51–52
   molecular genetics of, 52
Nephrogenic diabetes insipidus, 15
Nephrogenic syndrome, 24
Nephrolithiasis, 53
Neurohypophyseal diabetes insipidus, 16
Neuronal migration abnormality, 5
NF-κB factor, 9, 34
NH2-terminal signal peptide, 34
Nitric oxide (NO), 138
N-linked glycosylation, 3, 37–38
Nonpeptide vasopressin receptor antagonists, 24
Nonsteroidal anti-inflammatory drugs, 150, 152–153
NPS 2143, modulator, 36
NPS R-568, modulator, 43
NSAIDs. *See* Nonsteroidal anti-inflammatory drugs
NSHPT. *See* Neonatal severe hyperparathyroidism

O

Optic atrophy, 16
Orthosteric agonists, 35
Oxytocin, 16

P

Pachygyria, 5
Palmitoylation–isoprenylation (CAAX) sequence, 142
P823A mutation, 43
Parathyroid hormone, 32–33

Parathyroid hormone-related peptide (PTHrP), 72
Parathyroid hyperplasia, 49
P179A (TMV) mutation, 141
Peptide linker, 41–42
Phenylalkylamine compounds, 36
Phosphatidyl inositide (PI), 66
Phosphoinositol, 42
Phospholipase C (PLC), 44, 142
Phospholipase D (PLD), 46
PHPT phenotype, 70
PKA phosphorylation, 143
PKDREJ protein, 3
Platelet-activating factor (PAF), 146
PLC activation, 143
Polyadenylation signal sequences, 34
Polycations, 35
Polycystic kidney disease protein 1 (PDK-1), 3
Polydipsia, 20
Polymorphisms, 66, 116, 122, 155
 in CASR, 56
Polyuria, 20
Polyuro-polydipsic syndrome, 16
Progesterone, 99
Proinflammatory cytokines, 73
Proline residue (P17), 140
Prostacyclin
 agonists, 138
 based therapies, 134
 binding, critical residues involve in, 141
 discovery, 134–137
 induced antiproliferative effects in VSMC, 147–148
 inhibition of VSMC migration, 148
 PPAR signaling, 139
 in promoting VSMC differentiation, 146
 receptor
  ligand-binding pocket, 139–140
  secondary structure, 139
 signal transduction, 140
Prostaglandin E2, 20
Prostaglandin X (PGX), 134
Protein kinase A (PKA), 17
PTGIR gene, 138
PTH. *See* Parathyroid hormone
PTH gene, 52
PTH secretion, 66, 73
Pulmonary hypertension, 154

R

R212C phenotype, 155
Receptor-activity-modifying proteins (RAMPs), 45
Rectifying potassium channel (ROMK), 54
Renal stone, 20
R66H and R66C mutations, 57
R137H mutant, 24–25
Rho-dependent transcription activation, 9
Rhodopsin, 2, 17, 120

S

SCH66336 inhibitor, 142
Secondary hyperparathyroidism, 73
Secretin receptor (class B), 2
Seizures, 33
Selective COX-2 inhibition, 152–153
Sensipar, 73
Sepsis, 73
Serines, and receptor regulation, 142–143
Seven-transmembrane (7-TM), 33
Single nucleotide polymorphisms (SNPs), 67, 154
Sodium–potassium–chloride cotransporter, 54
Spontaneous ovarian hyperstimulation syndrome (sOHSS), 121–123
Stat1/3 elements, 34
Stimulatory G protein (Gs), 17
suREJ3 protein, 3
S128Y mutation, 122

T

TATA box, 34
Testosterone, 98, 104
Thrombospondin, 2
Thromboxane A2 (TXA2), 134
Thyrotropin receptors, 117
Tissue transglutaminase (TG2), 8
T151M mutation, 59
7-TM receptors, 34
Transmembrane domain (TMD), 34, 42
TSH receptor (TSHR), 116
TSHR mutation, 121
Tunicamycin, 37
Type II LCH, 105

INDEX

U

3'-Untranslated region (UTR), 34

V

Vascular smooth muscle cell, 138. *See also* Prostacyclin
  differentiation, 146
  phenotypic modulation, 145–146
  proliferation, 145–146
  prostacyclin-induced antiproliferative effects in, 147
Vasopressin, cellular actions, 16–20
  in distal nephron, 20
  to increase water permeability, 18
Vasopressin-sensitive water channel, 16
Vasopressin V2 receptor. *See* V_2 receptor
Venus-flytrap-like domain (VFT), 34, 37
Vitamin D response elements (VDREs), 34
Voltage-gated chloride channel, 54

Voltage-sensitive calcium channel blockers, 36
Vomeronasal sensory organ (VNO), 34
V_2 receptor, 15, 17–19, 24
V2R-specific effect, 23
VSMC. *See* Vascular smooth muscle cell

W

Walker–Warburg syndrome (WWS), 5
Wolfram syndrome 1, 16
W71X mutation, 21

X

X-linked recessive NDI, 15

Z

Z mutation, in α1-antitrypsin deficiency, 23

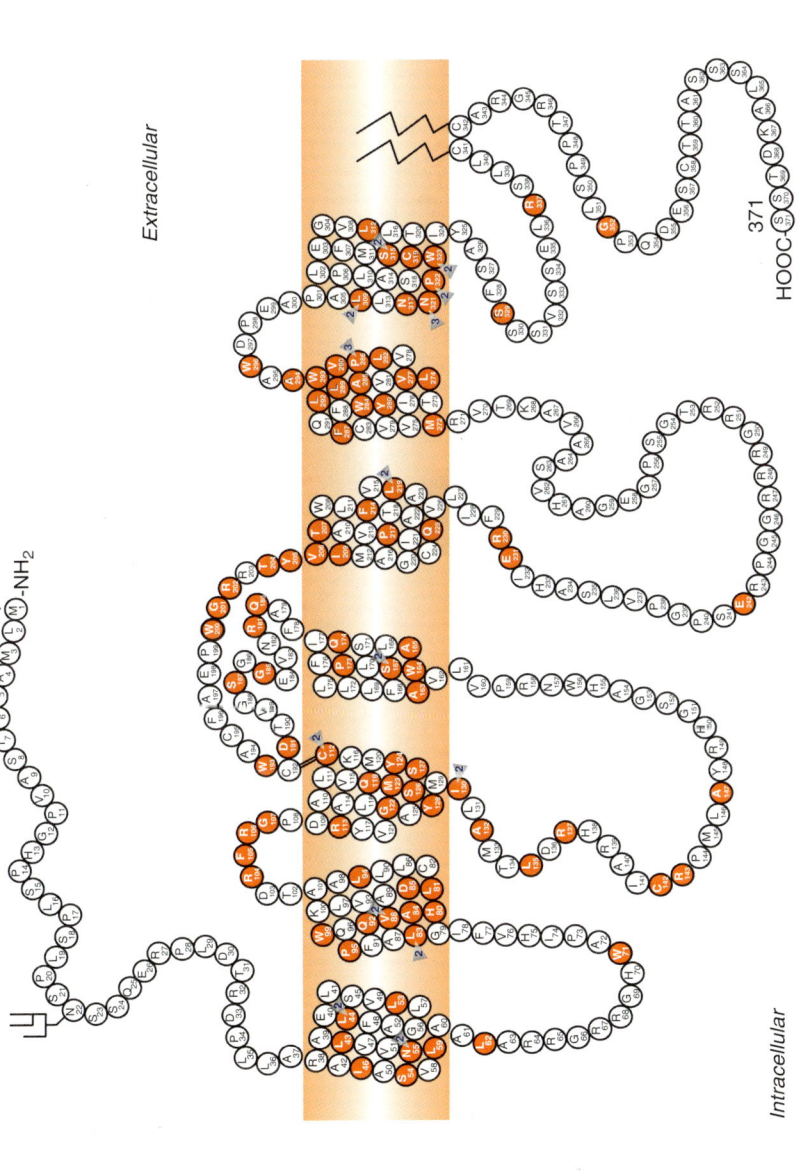

DANIEL G. BICHET, FIG. 2. Schematic representation of the V_2 receptor and identification of 193 putative disease-causing *AVPR2* mutations. Predicted amino acids are shown as the one-letter amino acid code. A solid symbol indicates a codon with a missense or nonsense mutation; a number indicates more than one mutation in the same codon; other types of mutations are not indicated in the figure. The extracellular, transmembrane, and cytoplasmic domains are defined according to Mouillac et al.[66] There are 95 missense, 18 nonsense, 46 frameshift deletion or insertion, 7 in-frame deletion or insertion, 4 splice-site, and 22 large deletion mutations, and 1 complex mutation.

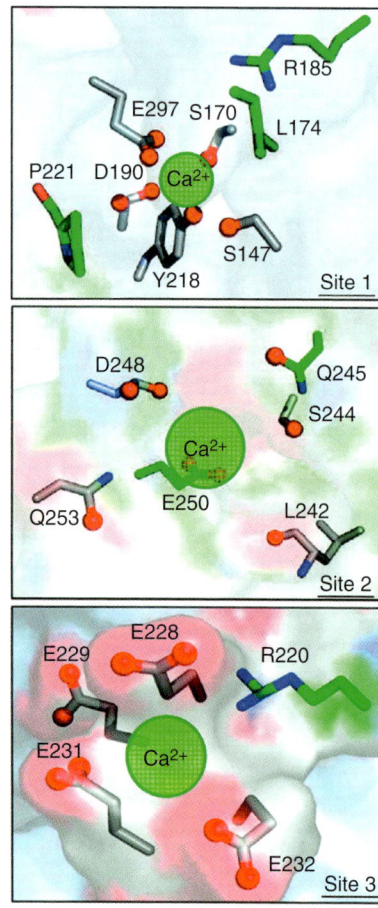

HENDY ET AL., FIG. 4. Location of three predicted Ca^{2+}-binding sites in a subdomain (amino acids 132–300) of the ECD of the CASR. Site 1 is in the hinge region of the VFT structure formed by lobes 1 and 2 (see Fig. 1). Sites 2 and 3 are in the first half of lobe 2 (amino acids 215–253). Sites 4 and 5 (not shown) are clustered in the second half of lobe 1 (amino acids 350–400). (Redrawn with permission from Ref. 95.)

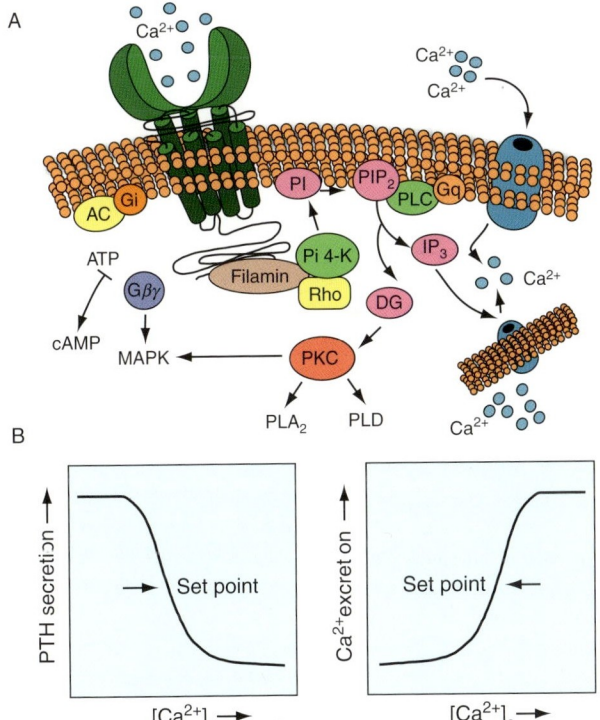

HENDY ET AL., FIG. 5. (A) Signaling pathways activated by the CASR. (B) The CASR controls the relationship between the extracellular calcium concentration $[Ca^{2+}]_o$ and PTH secretion on the one hand and urinary calcium excretion on the other. The set-point is the $[Ca^{2+}]_o$ at which PTH secretion or calcium excretion is half-maximal.

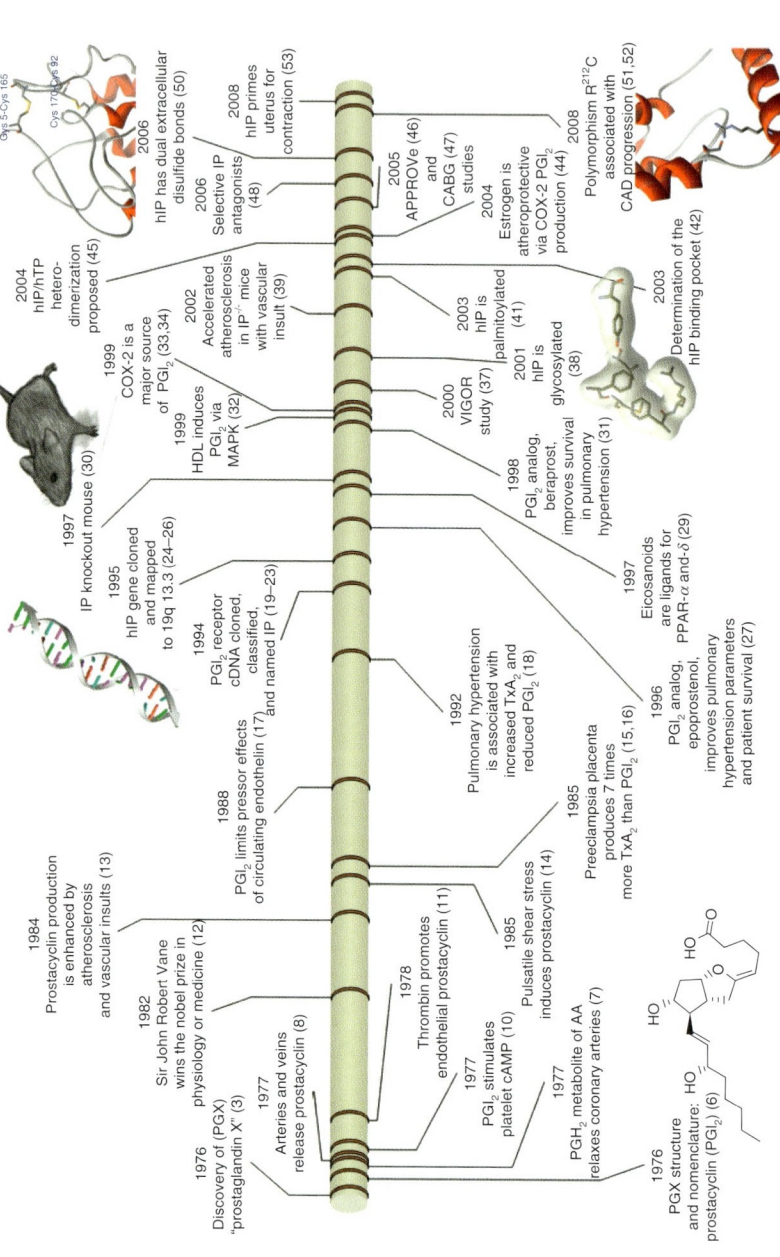

MARTIN ET AL., FIG. 1. Timeline describing a brief history of some of the major landmarks in the study of prostacyclin from its discovery by Vane and his colleagues to recent interesting insights into roles played in normal physiology and pathology. It can be seen that the initial series of foundation studies were followed by a period of relative quiescence until the gene (PTGIR) was cloned, knockout mice studies performed, and COX-2 inhibition appreciated. The increasing interest in genetic variants will likely lead to many new insights into the role played by the prostacyclin receptor in human pathophysiology.